Graphics HANDBOOK

A BEGINNER'S GUIDE TO DESIGN, COPY FITTING AND PRINTING PROCEDURES

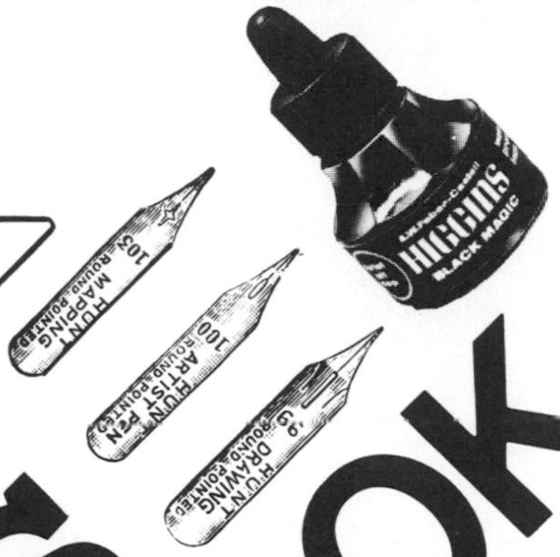

by
HOWARD
MUNCE

Graphics HANDBOOK

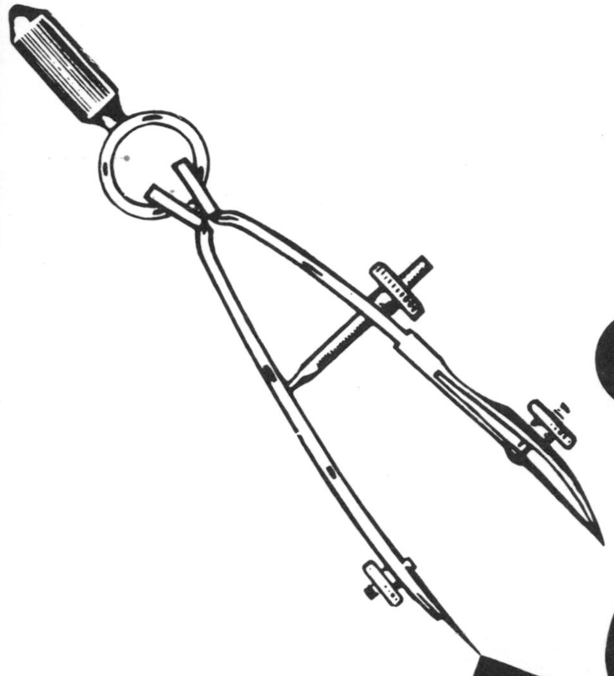

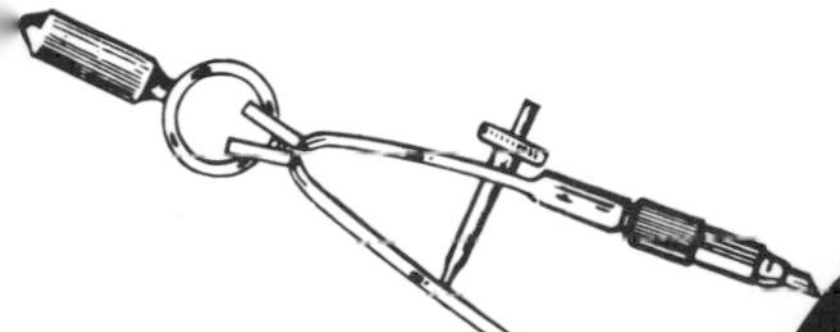

by HOWARD MUNCE

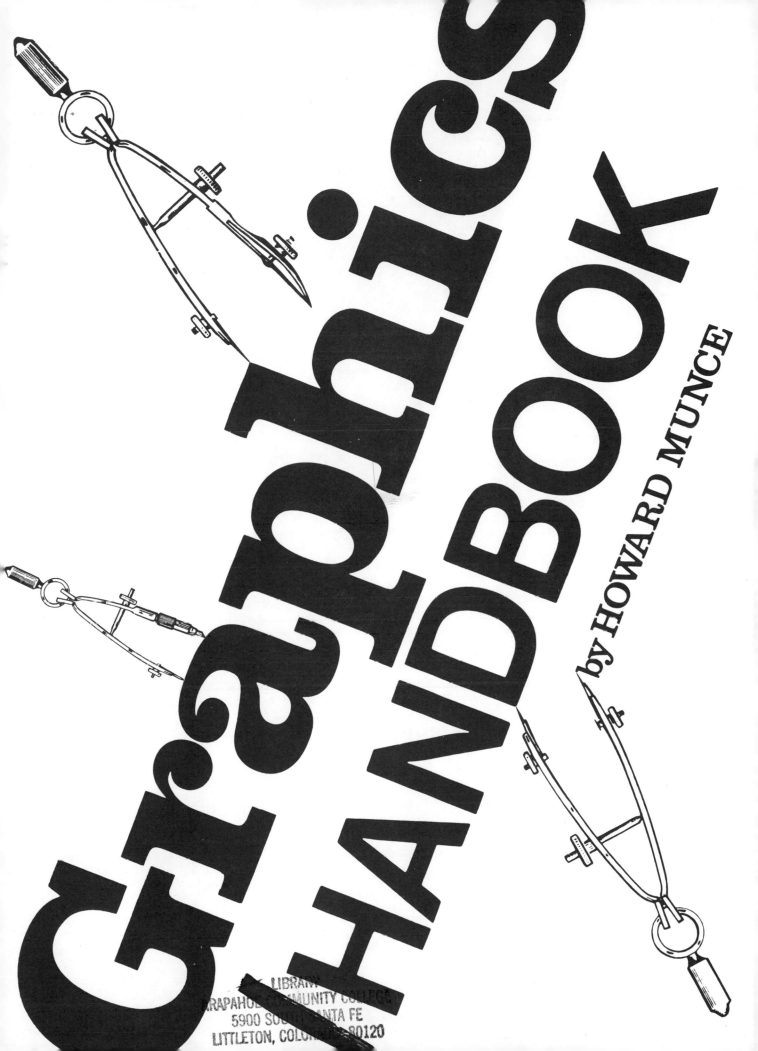

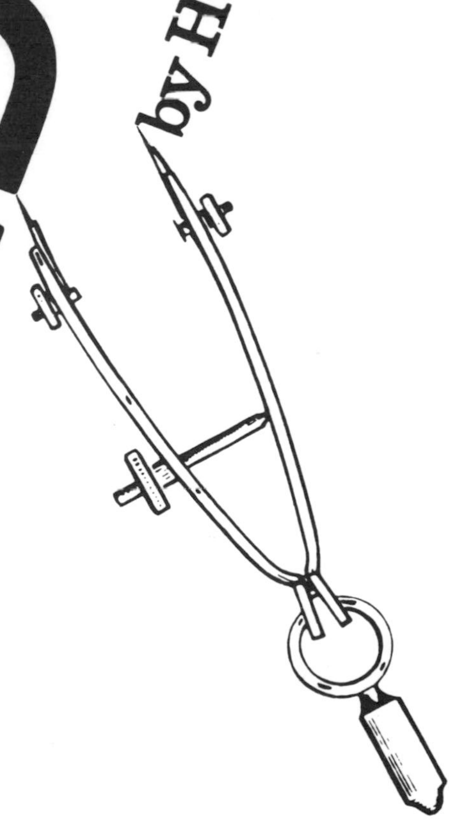

Published by NORTH LIGHT PUBLISHERS, an imprint of
WRITER'S DIGEST BOOKS, 9933 Alliance Road,
Cincinnati, Ohio 45242.

Manufactured in U.S.A.
First Printing 1982
Second Printing 1983
Third Printing 1984
Fourth Printing 1986

[redacted]

Library of Congress Cataloging in Publication Data

Munce, Howard.
Graphics handbook.

1. Printing, Practical—Handbooks, manuals, etc.
I. Title.
Z244.5.M85 1982 686.2 82.8278
ISBN 0-89134-049-1 (pbk.)

Edited by Carla Dennis
Designed by Howard Munce

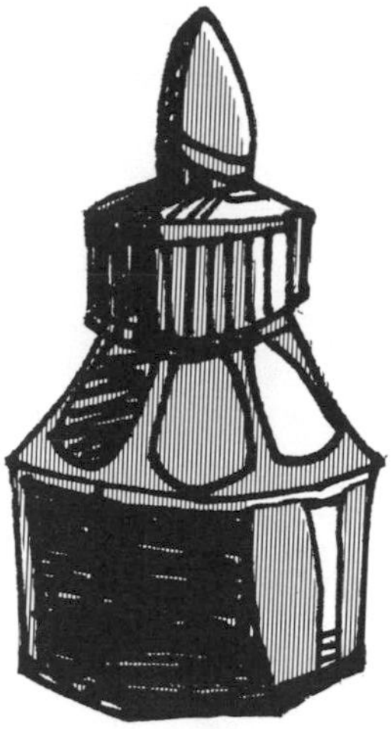

CONTENTS

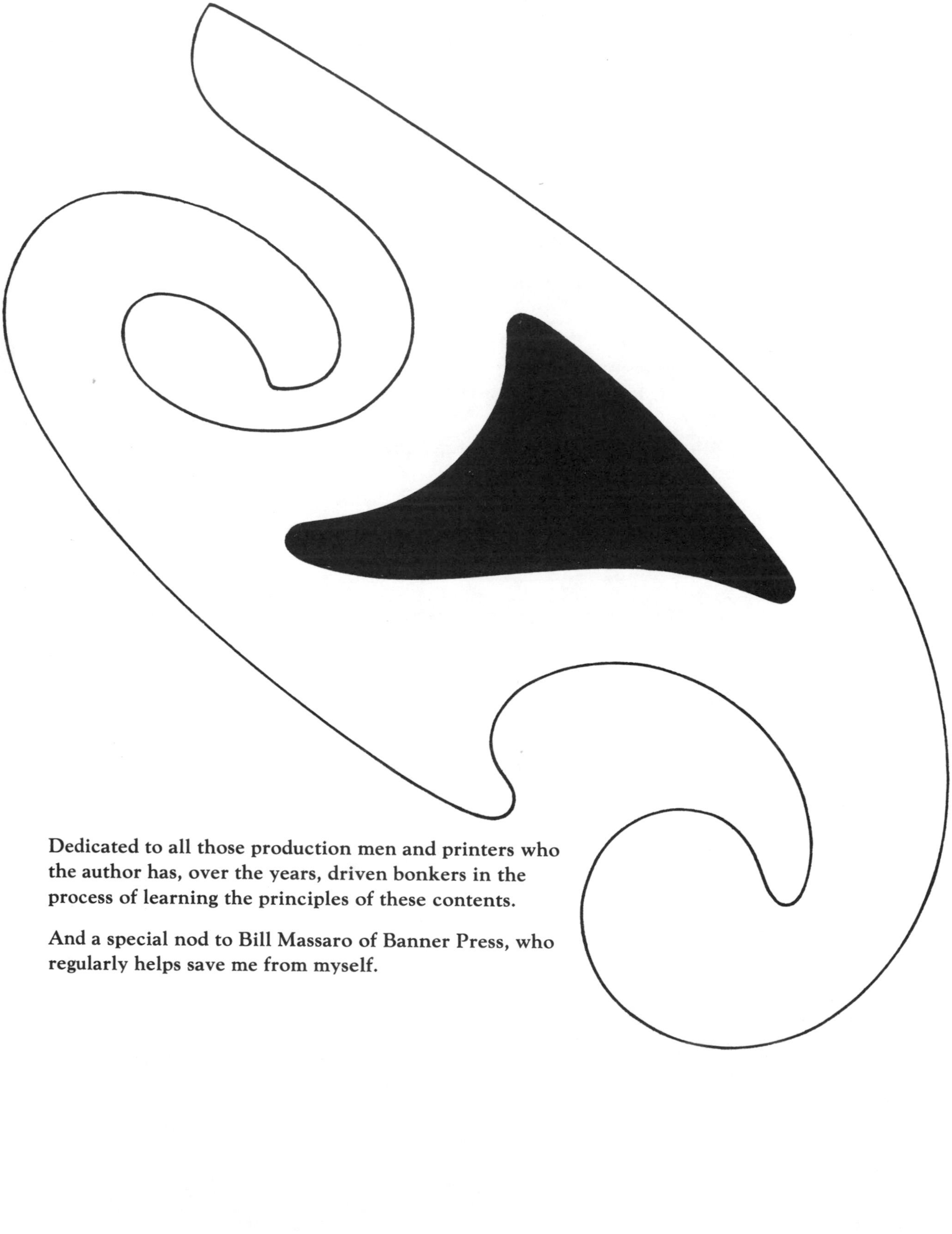

Dedicated to all those production men and printers who the author has, over the years, driven bonkers in the process of learning the principles of these contents.

And a special nod to Bill Massaro of Banner Press, who regularly helps save me from myself.

INTRODUCTION

This book has a single aim: to take the fright, bewilderment and mystery out of the design and preparation of simple printed pieces.

It has been prepared for those people who suddenly find themselves appointed to be in charge of mailings, brochures, flyers, announcements, booklets and advertisements. Also for those who wish to design their own greeting and commemorative cards.

There is a need for such material in all phases of community life—for your clubs, libraries, churches, political parties, charity groups, scout troops, cultural meetings, and on and on. Take a look at a crowded bulletin board and see for yourself.

Beyond these, there are other opportunities. I have conducted workshops on the subject, where people with an interest in graphics but with no training have come to learn, because there are opportunities in their business to further themselves if they can add this function to their other duties.

Also, many merchants have a growing interest—they either want to learn to understand graphics themselves, or they want to familiarize themselves enough to be knowledgeable about what their advertising people produce for them.

There is also an interest among many young people just to grow in graphics because it's an exciting part of their generation—they've been exposed to everything from wild record album covers to posters to mad tee shirts. They're aware of printed things and are alert to things having a "look."

I'd like you to consider this book as a workshop session between you and me. We'll go about it at a nice simple pace. I have tried to anticipate your questions and build the answers into the text, the captions and the illustrations.

You'll be surprised at how easy the mechanics are. And I hope I can whet your appetite enough to make you want to become an effective designer with respect for letter forms, legibility, simplicity, style and taste.

Pull up a chair—we'll talk.

Glossary

Advertising Agency
An organization of artists, writers and other specialists equipped to create and produce advertising.

Airbrush
A pen-sized instrument used for applying a fine spray.

Art Director
One who directs the art policies of an advertising agency or publication. A person familiar with art styles and reproduction. One who creates visual ideas and commissions illustrators or photographers to carry them out.

Art Service
An organization of artists of varied specialties.

Benday
An engraving process wherein solids are broken into areas of dots.

Bleed
Printing art work or a photo without leaving any margin between the picture and the edge of the page.

Blowup
An enlargement of a drawing or photo.

Blueprint
A proof made for the purpose of checking all elements and final proofreading, before the printing plate is made.

Body Type
Type set for the main body of printed matter.

Boldface
It is heavier and darker than light or medium type.

Booklet
A paper covered pamphlet.

Brochure
A stapled pamphlet.

Calligraphy
Flourished writing or penmanship.

Camera-Ready
That which you deliver to the engraver or printer—the mechanical and all art elements that apply to the job.

Caption
Descriptive lines of copy for an illustration.

Caricature
A comical exaggeration of the characteristics of a person.

Cartoon
A comic or satiric drawing.

Casting
Selecting people as models.

Character
An individual letter, number or punctuation mark—all to be reckoned with in copy fitting.

Coated Paper
Paper which produces a fine, smooth surface. Also known as coated stock.

Collage
The arrangement and affixing of materials on a surface to form a composition.

Compass
An instrument for drawing circles.

Contrast
Strong differences or opposites in form, line, texture or color.

Copy
The text to be used for printing.

Copy Fitting
Specifications for the setting of text. Selecting the typeface and specifying its size, the width it is to be set, and all other instructions for the typographer.

Crop
The elimination of unwanted areas.

Crop Marks
Horizontal and vertical marks to indicate limits of art.

Deckle Edge
Rough or uneven edge of some printing papers.

Decorative
A style that takes merry liberties with realism.

Descenders
The parts of lower-case letters that drop below the main body. Ascenders rise above the main copy.

Design
A planned arrangement of the elements in a composition.

Display Type
Type larger than text size for headlines and subheads.

Dividers
The two-pronged instrument used for dividing lines into equal spaces or for transferring measurements.

Dummy
Mock-up of printed piece, in its sketched or pasted-up form, to indicate final appearance.

Ellipse
A circle in perspective.

Engraving
The process of transferring an image to a metal printing plate by mechanical means.

Face
Style and name of typeface.

Finished Art
Art work or photo—ready for reproduction.

Flop
To reverse.

Folio
A sheet of paper folded once—which produces four pages. Also page numbers in a publication.

Glossy Print
Shiny-surfaced photo print.

Gothic
One of a group of typefaces.

Gutter
The inner margin of a printed page extending from the printed portion to the fold or binding.

Halftone
A process for reproducing tonal images by photographing copy through a screen.

Illustration
A picture that interprets a written piece. A pictorial interpretation of a situation or of an idea.

Illustration Board
Drawing paper mounted on cardboard, used to make drawings or to do mechanicals.

Italics
Type slanted to the right.

Job Printer
One who does miscellaneous printing, such as circulars, cards, stationery.

Layout
An art director's or designer's sketch of how a printed piece will appear. Done for client approval and as guide for illustrator or photographer.

Leaflet
An unbound single sheet, moderate in size. It may be folded or left flat.

Letterhead
The printed matter on a sheet of letter paper.

Lettering
Hand executed words.

Line Drawing
A drawing composed entirely of lines, dots and areas of solid black. It can be reproduced by line process.

Logotype
The lettered signature name plate or trademark of a commercial firm. Sometimes called a logo or slug.

Lower Case
Letters of the alphabet uncapitalized.

Measure
Width of a column or line of type.

Mechanical
A paste-up and working guide for the printer or engraver. All the elements are pasted or indicated in their correct positions.

Offset Printing
The printing process in which impressions are made on a rubber blanket by a metal plate, and are then transferred to paper.

Overlay
A transparent sheet taped over the art to indicate the parts of the drawing to be printed in color, or any other instructions to be followed.

Pamphlet
A paper bound booklet with or without a separate cover.

Paste-up
See **mechanical.**

Photolettering
Alphabets done photographically.

Photostat
Image recorded by a camera that photographs and develops directly on paper while you wait.

Phototype
Type set by photographic means.

Pica
Printer's unit of measure. Each pica is one-sixth of an inch, or to put it the other way around, there are 6 picas in an inch.

Register
Exact fitting together of two or more elements to be printed together.

Reproduction
Printing one or more copies of an original piece of work.

Rescale
To enlarge or reduce the size of a picture without changing the proportions. To refit elements into new size or shape.

Research File
A collection of pictures used as reference. Also known as a scrap file, swipe file, clip file, or morgue.

Retouching
The alteration or repair of details in a picture; the removal of spots and blemishes on a photograph.

Reverse
To turn over, to make the right-hand side the new left-hand side. See **flop.**

Roman
Alphabets or typefaces characterized by thick and thin strokes and serifs.

Rubber Cement
A flowing adhesive. Also comes in spray cans.

Ruling Pen
An instrument that holds ink between two adjustable prongs to allow for different weights of line.

Sans Serif
Letters or type without serifs.

Screen
A glass plate or film with crosshatched lines through which art work is photographed, producing dots of varying size.

Script
Lettering resembling handwriting.

Serif
Horizontal parts of letters at top and bottom.

Silhouette
The solid flat shape of a form. The isolation of a form by removing the background.

Small Caps
Capital letters smaller than standard capital letters of a typeface.

Triangle
Triangular-shaped ruling guide, available in several sizes and angles. Made of plastic or metal.

T-square
A ruling guide that is held against the straight side of a drawing board, primarily to make horizontal lines. They're made of wood, aluminum or steel.

Typeface
A particular type design, sometimes designated by the name of the designer—but always with a name of some kind. Often the same face is known by different names.

Typography
Art of type selection, arrangement, fitting and setting.

Upper Case
The capital letters of the alphabet.

Zip-A-Tone
A transparent waxed sheet covered with a variety of patterns, dots or lines. They are placed over line art to produce tonal effects. Commonly used by newspaper cartoonists and chart makers. Reproduction is cheaper with this method. Manufactured by many companies—all under different trade names.

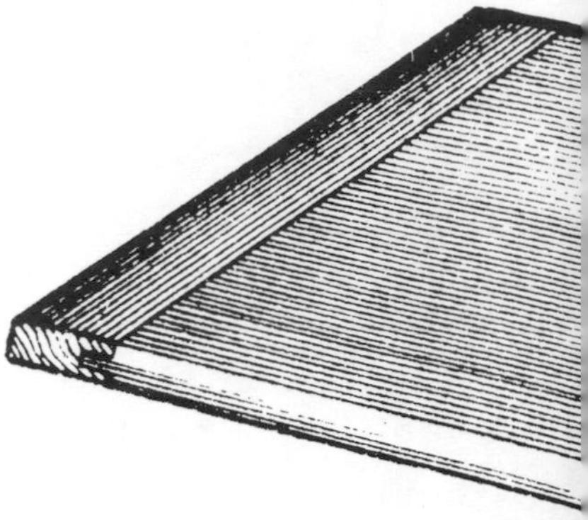

MATERIALS:
A showing & description of the few basic tools needed in the preparation of the printed page.

Working in the graphic arts field has one big advantage over other kinds of art work—and that's in the materials department.

The tools of the designer are fewer, less messy and easier to move from place to place. And though they're not cheap, they will, with care, last indefinitely.

Assuming you're starting from scratch, this is what you'll need to acquire.

The drawing-board or table: These come in several styles and sizes. Many artists prefer a small basic one with the option of attaching a larger board to it. Your space and purse will help you determine which you'll buy. If you can afford the kind of table that can be adjusted easily to allow you to stand up to work when you wish, by all means get it. Not only does this type allow you to position yourself correctly for certain situations, but it also affords great relief to your neck, back and bottom when you've worked seated for long stretches.

If your board doesn't come with an attached metal edge on the sides, you'll need to get one that clamps on. These are imperative for accurate work.

The T-square: This is another item that comes in many forms. But it's important that you begin with a *good steel* one. I suggest the 26-inch one.

The T-square will become almost a part of you. Treat it well—next to a broken watch I can't think of anything more useless than an inaccurate T-square.

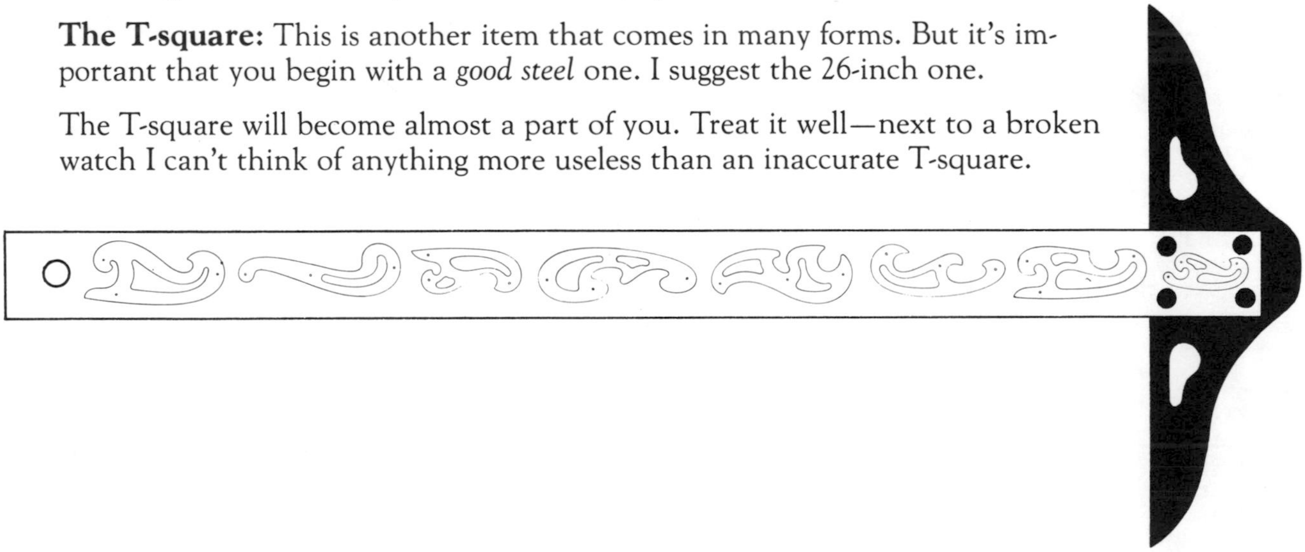

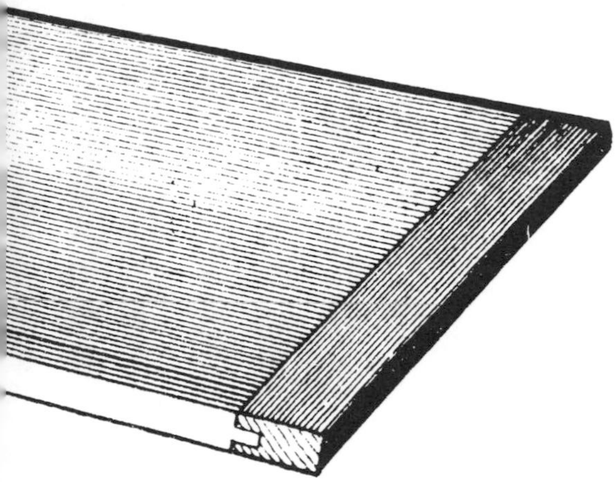

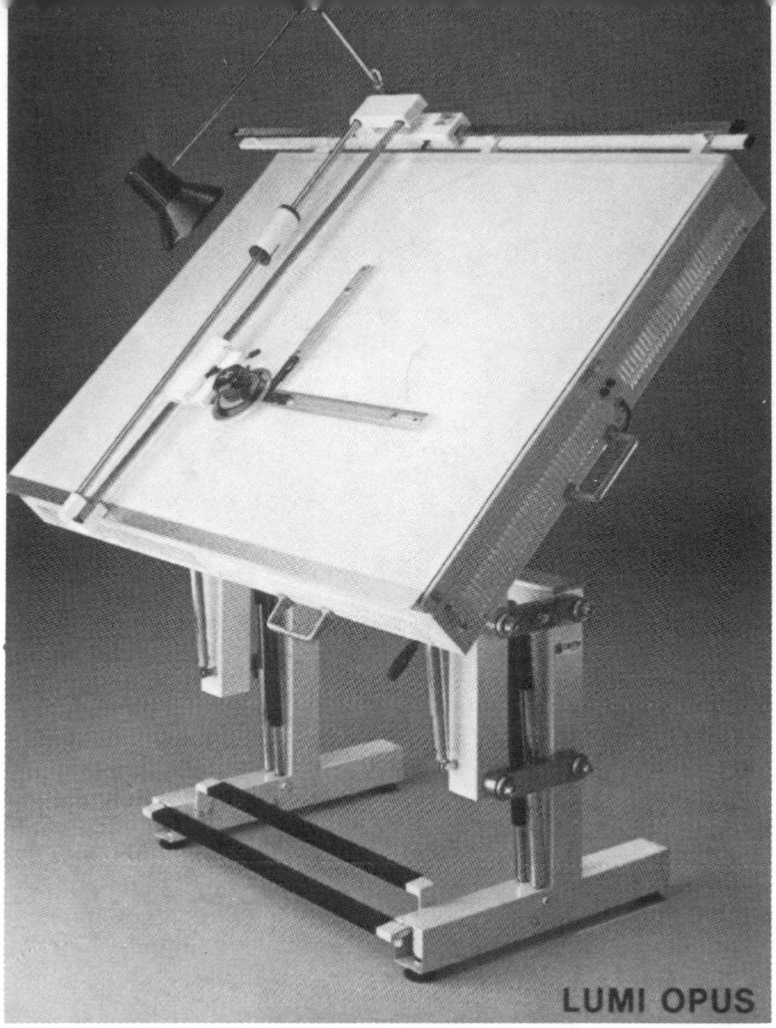

The two drawing boards above range from the trusty old breadboard type pine slab to the super Space Age beauty on the right—the Lumi Opus by Bieff, imported by Sam Flax of New York and Atlanta. There are dozens of models in between—and more on the drawing board, so to speak. Take your choice.

Most people put their straight edges on the left or right side of their boards, depending on whether they're left or right-handed. You'll handle the T-square with the opposite hand from the one you draw with. In either case, such an arrangement will give you accurate horizontal lines. Of course, you'll need the same accuracy for your vertical lines. Unhappily, manipulating your T-square from the top or bottom of your board is not likely to be accurate enough. This is where your triangle must be used. Again, they come in many variations and sizes. Eventually you'll acquire most of the ones you need. Their basic use is to give you a perfect angle. To accomplish this, you merely set the base of the angle along your T-square and glide it left or right.

Keep your triangles clean and *don't* use them as a cutting edge. One nick and they're ruined for ruling. To use them for ink ruling, it's necessary to put a narrow strip of masking tape just back from the edge. This will elevate it enough to keep the ink from spreading underneath. Tender your T-square and all other tools you use for ink ruling the same gentle treatment. They'll drive you mad if you don't!

Welcome newcomer: *this will be your playing field and your battlefield. You'll have joyful times working on this area—and depressing moments—just know that with practice you'll improve rapidly and be greatly rewarded by your sense of accomplishment.*

Here you see your most basic tools laid out on a clear and pristine board; some of you will always keep your work surface as immaculate as this. Others, more like the author, will soon have it marked up with cut scars, ink spots, telephone numbers and doodles. Neat is best.

On the right is a set of mechanical drawing tools: ink and pencil compasses, a pair of dividers used to repeat spaces and areas and an ink ruling pen. You'll need one of each.

On the next page is a selection of French curves—you won't need all of them of course—but it's wise to keep adding new shapes to your collection. The same is true of the ellipse (a circle in perspective) templates. They range from 10° to 80°. Indispensable for sharp and accurate work.

Below is rubber cement shown in its many-sized containers. I recommend what is called "one-coat." Logically enough, it's half the work of "two-coat."

Rubber adhesive in spray cans is efficient—especially if you don't need to use cement frequently. The spray cans don't dry up as other containers tend to. Commercial studios and large art departments now mostly use hot wax; excellent, but it doesn't make sense unless used regularly. The wax machine is costly.

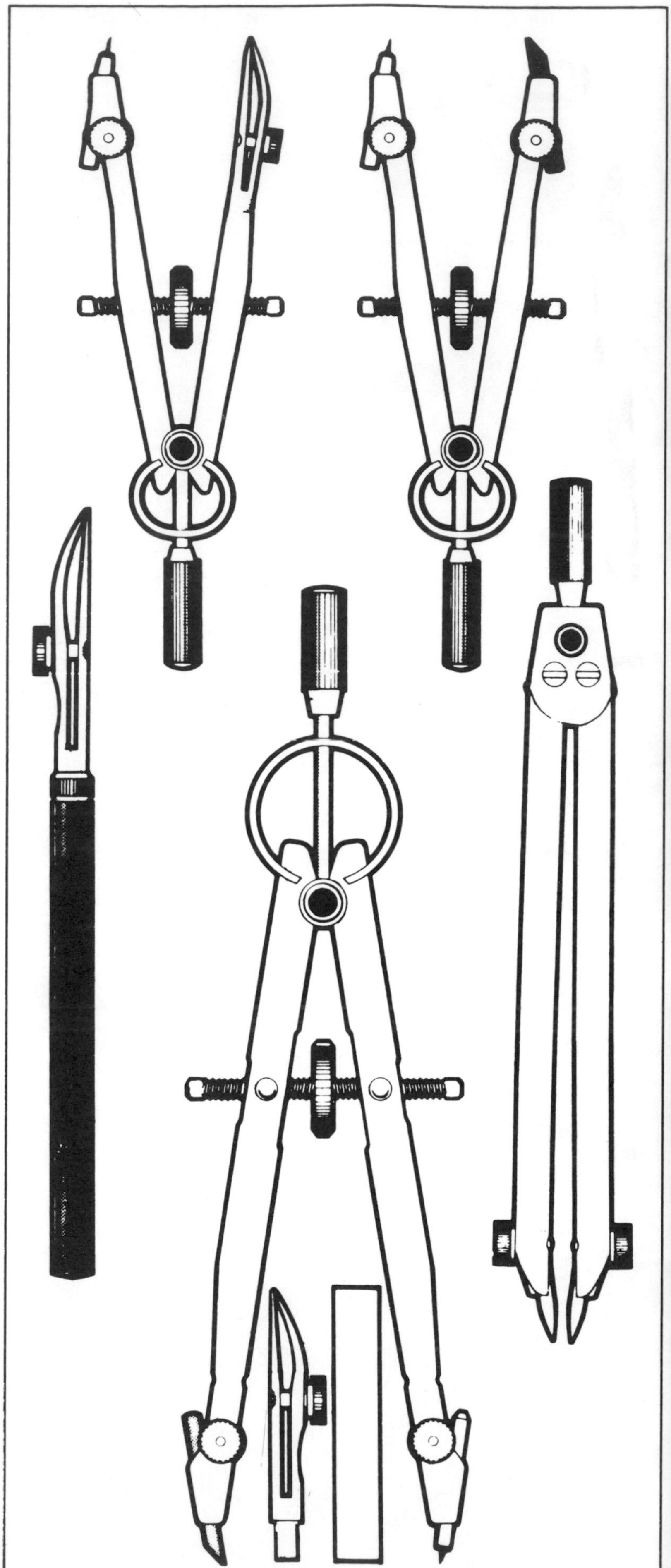

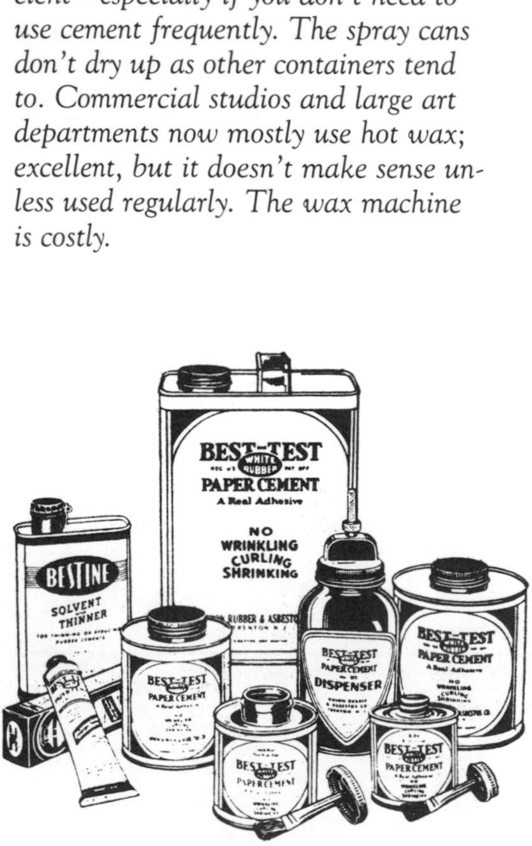

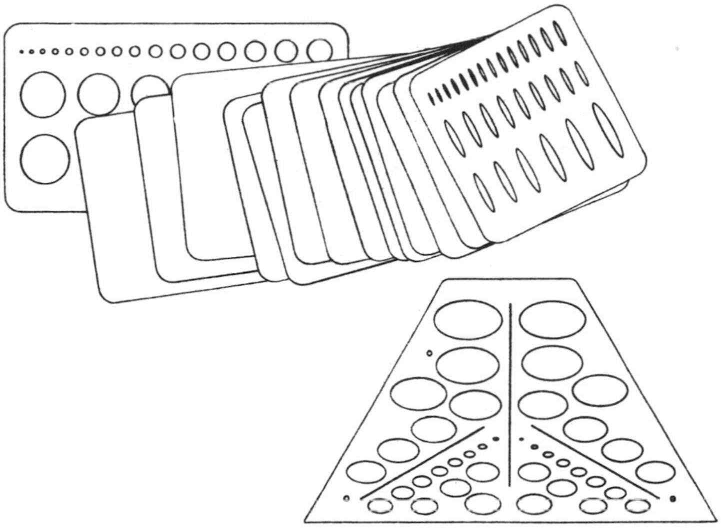

Here are some further necessary implements for your graphic tool kit. They are all basic things—but I warn you, once you get truly involved in this pursuit, you'll be hooked, and you'll find it difficult to ever again leave an art supply store empty-handed. Welcome to the club.

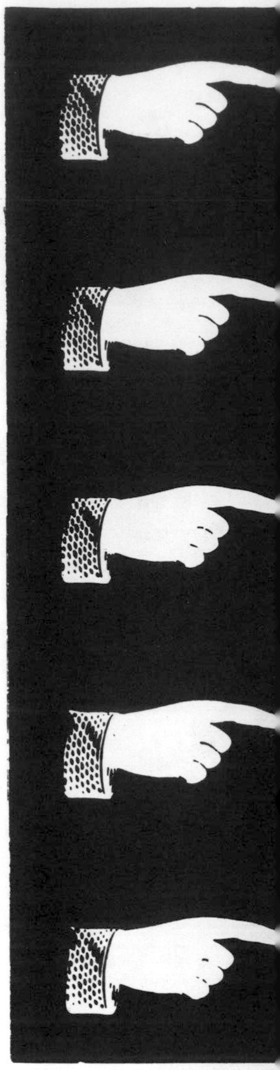

COLLECTING A TYPE FILE:

Several examples of the kinds of graphic oddments you should stash in your file for inspiration & emergency.

Most artists and designers are "savers." We compulsively cling to things because we like so many odd items that the average person doesn't notice. We hold on to stuff because we can use it for inspiration, or in some cases we actually *re-use* it—like numerals from obsolete calendars, for instance, or bits of decorative gift wrappings that we can put to other design or collage purposes.

This is not to suggest that you should plagiarize, for that's despicable conduct for anyone who regards himself as creative. I'm referring only to odds and ends of type, ornaments, dingbats, old engravings, borders, punctuation marks and other graphic devices that you can incorporate into printed assignments, either for spice or to rescue you from a bad spot when your type runs short and you have a gap to fill.

Also, learn to save examples of printed pieces that have impressed and moved you. Without lifting their ideas, peruse them every so often. Ask yourself if you've tried to be as daring—or original or as tasteful. If you've unconsciously fallen into the rut of doing everything safely and conventionally, an occasional browse in your "good file" will jar you out of your complacency and make you take stock of yourself. This process of self-evaluation and goading yourself to better things goes on forever in an artist's life. Look around you at the superb work that exists—and keeps coming. In this way you'll see what the professional standards are.

Here's a simple file system I recommend—(fashion yours to fit your own needs):

I have a folder marked "Numerals" that contains collected numbers from calendars, printed pieces that came in the mail, or were found in magazine articles and from old jobs of my own.

The folder marked "Letters" contains examples of anything that looks like it might serve a purpose later on—usually as an initial capital letter.

The folders for borders, ornaments, old engravings and punctuation marks will obviously be labeled as such.

I have a folder for ampersands (&) alone. They are always a useful tool for the designer, for not only do they have a unique graphic *look*, but they allow you to substitute for the word *and*—this is especially useful in logotype design where that word occurs often.

Enlarged quotation marks are another useful device. They allow you to give emphasis to quotations in ingenious ways.

Keep a folder, too, on pointing fingers, which exist in all varieties. Some very early ones are black silhouettes. Some of the old ones are crudely drawn—others are beautifully engraved or come from carved wood type. Each has a distinct look. There are many more contemporary versions of this useful device, including the ones now available on transfer sheets.

Arrows should be saved. They, too, come in a never-ending variety of styles, weights, delicacy, bluntness and lengths. There are also curved ones—hence they can do what the pointing fingers *can't do*—point around corners. Bear in mind that you'll need *rights* and *lefts* in hands and in some arrows.

I've found it a good practice to examine my leftover type from every job to see if there is anything worth saving. Often there is—especially in the display-size letters, numerals and punctuation marks. All of these can be resuscitated by enlargement for later use.

Also, get in the habit of cementing or taping on some of your favorite type devices to other work that you're having photostated. This way they get a free ride, and you'll get the pieces back in a new size each time, depending on the focus of the original thing you're having stated.

Lastly, here's a very important folder to keep—especially if you use the same type face often, as many people do with the popular Helveticas, for instance. Hold on to extra type proofs from each job and note the name, size and weight on each piece. One day—or especially one *night* or weekend when new type is unavailable, you'll be glad you did, because if you have the need to make a correction or patch a word, you'll be able to do so. Such small type errors are referred to as *typos*.

Also save all extra sets of folio numerals (page numbers). Often you'll be able to use a whole set intact—or at least you can patch if the type face is the same.

Eventually your type file will take on a special flavor: it will reflect you and your particular tastes—and that's how it should be.

...2 2
5
2
4
1 2 3
1 2 3 4 5 6
9 10 11 12 13
17 18 19 20 21
25 26 27 28 29
$
1
2
3
4 5

½ ⅓ ⅔ ¼ ¾ ⅛ ⅜ ⅝ ⅞

⑥ ⑦ ⑧ ⑨

34 56

0

① ② ③ ④ ⑤ ㊀65 95 98 99

1 2 3 4 5 10 13 15

❶ ❷ ❸ ❹ ❺ ❻ ❼ ❽ ❾

6 ③④ ⑤⑥ 7 8 9 ⓪

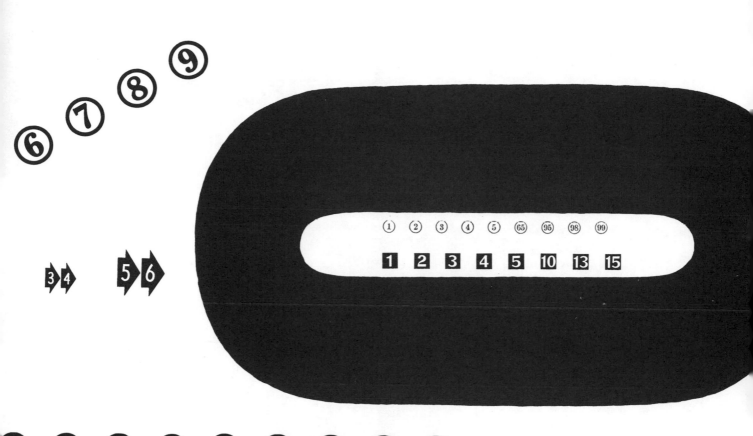

? ? ? ? ? ? $? ! $

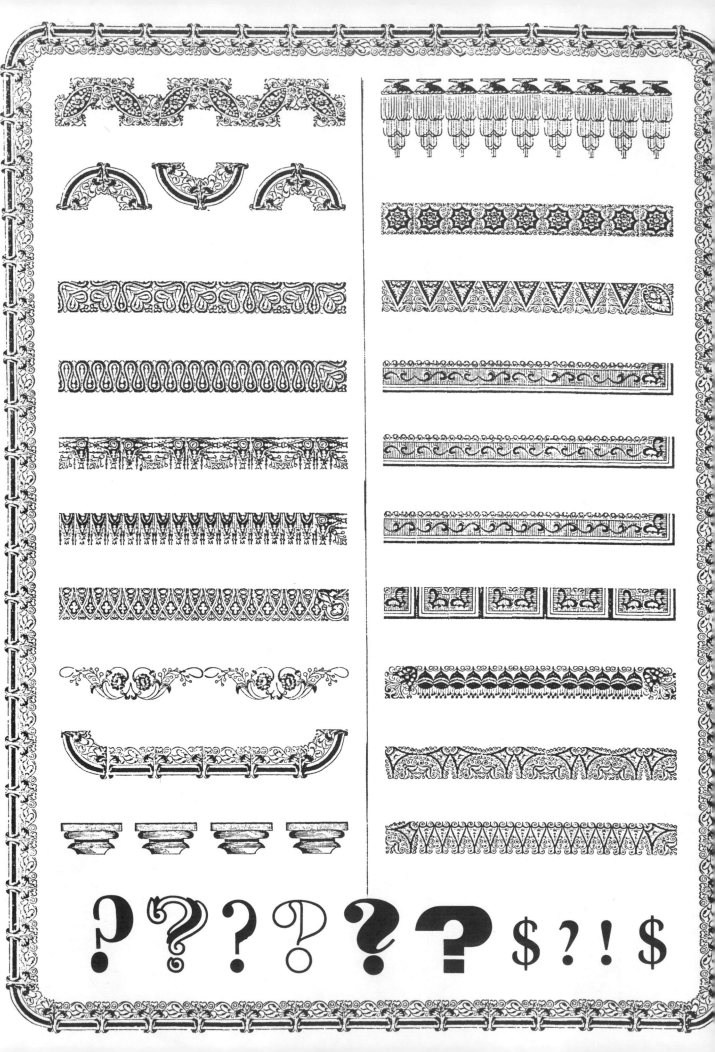

Did you realize that quotation marks, ampersands, et al., come in so many forms?

ABCDEFGHIJKLMNOPQR
STUVWXYZ 1234567890

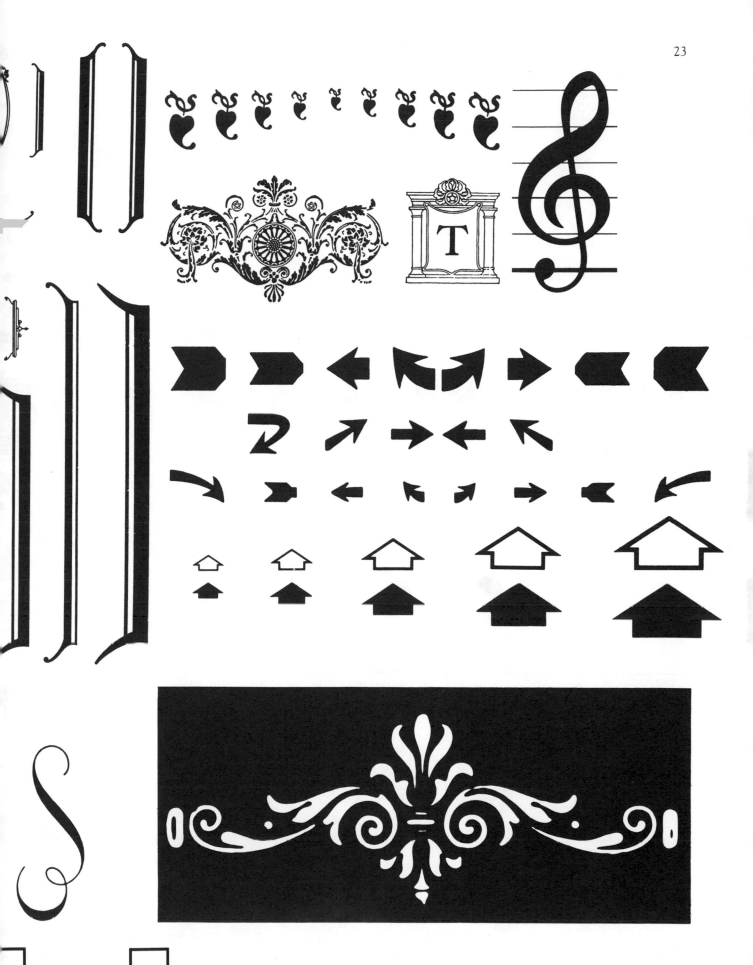

WATERCOLORS!

OM A ton

BCD E

FALK

Illust

om, THE E

T G X Y

Z

&

atör's

MADE IN NE

THE USE OF TYPOGRAPHY:

An introduction to what seems a bewildering & frightening subject, but really isn't if you don't cringe as the author was once wont to do.

The most complicated step in the preparation of a printed piece is dealing with the type. So plunge in, learn how much simpler it can be than you feared, and get it out of the way. But do this with the understanding that you will only get a working knowledge from these pages, and that your real proficiency will come from working on actual projects wherein you select, fit, order and handle real type. It's only then that you can study the subtle differences between what you visualized and how it looks in the hand.

Also know that the reward of an interest in letter forms and type use will depend on your becoming more aware of the endless parade of printed words that confront your every waking hour.

When you've gained more awareness you will endlessly evaluate what, until now, you've been blind to. The bulk of the printed matter we are everlastingly assaulted with is tasteless trash. Much of it is sign painting—and most of its badness comes from advertisers trying to outshout each other by sheer size and garishness.

Look around you, and begin deciding what you think has merit and what offends. The good things are always: Legible, Inviting to read, Untricky, Simple. Become a critic—it's a good way to school yourself.

To begin this study you must arm yourself with a typebook. If you do enough business with a company that sets type they will give you one—if not, you'll have to buy one. By hook or by crook, get one.

It is from this book that you will make the following decisions:

1. *Choice of typeface (style)*
2. *Choice of size*
3. *Choice of leading (space between lines)*
4. *Choice of italic (slanted version)*
5. *Choice of bold version (to be intermixed if available)*
6. *And it is from this book that you will determine what area your copy will occupy when set.*

It is also common practice to take copy to the typographer and have him tell you how much room is needed for a given amount of copy.

But I want you to learn for yourself—so get your own book.

This is a typical page from a typebook. Most show one typeface at a time in all its available sizes.

Observe that the left and right columns are the same face and the same size. The right column gives the illusion of being different because of the amount of added space between the lines. You can have as much space (air) as your judgement dictates.

Stet Shields Inc.
STRATFORD, CONN.

10

Century Expanded is a light face type that is found most useful by many who specify types. In general characteristics it resembles the type faces used in the text matter of the readers used in the primary grades of schools, hence it is easy to read in the mass. The individual

ABCDEFGHIJKLMNOPQRSTUVWX
abcdefghijklmnopqrstuvwxyz 12345

Set Solid—7 lines to inch

Century Expanded is a light face type that is found most useful by many who specify types. In general characteristics it resembles the type faces used in the text matter of the readers used in the primary grades of schools, hence it is

ABCDEFGHIJKLMNOPQRSTUVWX
abcdefghijklmnopqrstuvwxyz 67890$

2 pt. leaded

Letters to Inch—Caps 10, Lower Case 14

12

Century Expanded is a light face type that is found most useful by many who specify types. In general characteristics it resembles the type faces used in the text matter of the readers used in the primary grades of schools, hence it is

ABCDEFGHIJKLMNOPQRSTUVW
abcdefghijklmnopqrstuvwxyz 12345

Set Solid—6 lines to inch

Century Expanded is a light face type that is found most useful by many who specify types. In general characteristics it resembles the type faces used in the text matter of the readers used in the

ABCDEFGHIJKLMNOPQRSTUVW
abcdefghijklmnopqrstuvwxyz 67890$

2 pt. leaded

Letters to Inch—Caps 8, Lower Case 12

14

Century Expanded is a light face type that is found most useful by many who specify types. In general characteristics it resembles the type faces used in the text

ABCDEFGHIJKLMNOPQRS
abcdefghijklmnopqrstu 12345

Set Solid—5 lines to inch

Century Expanded is a light face type that is found most useful by many who specify types. In general characteristics it resembles

ABCDEFGHIJKLMNOPQRS
abcdefghijklmnopqrstu 67890$

3 pt. leaded

Letters to Inch—Caps 7, Lower Case 10

AB AB AB AB

ROMAN GOTHIC ITALIC SQUARE SERIF

There are four basic kinds of letters: Roman, with serifs; Gothic, without serifs; Italic—letters that lean to the right; Scripts—a kind of written letter.

AB AB AB

BRUSH CARTOON SCRIPT

To give you just the slightest idea of some of the myriad typefaces, weights and variety that await your taste and imagination, I've included a few selections from the catalog of Central Typesetting Service, Inc., Memphis, Tennessee.

SERIF

ALBERTUS LIGHT	**Cheltenham Bold**	**Souvenir Demi**
ALBERTUS	***Cloister Black***	***Souvenir Demi***
ALBERTUS REG.	**COOPER**	Trooper Light
Americana	**DeVinne**	**Trooper Medium**
Americana Condensed	GOUDY OLD STYLE	**Trooper Bold**
Americana Bold Cond.	Melior	Tiffany Light
Arrow	Palatino	Tiffany Medium
Aster	*Palatino*	**Tiffany Demi B**
Baskerville	Serif Gothic	**Tiffany X-Bol**
Bookman	Souvenir Light	Trump Medieval
Bookman Italic	*Souvenir Light*	**Ultra Bodoni**
Caslon #123 Light	**Souvenir Medium**	**Windsor**
Caslon #123 Bold	***Souvenir Medium***	Windsor

SANS SERIF

rotesque	News Goth. Bold	SPIGOT
otesque	News Goth. Bold Cond.	SPIGOT
rotesque	News Goth. Ext.	SPIGOT
rotesque	News Goth. Bold Ext.	Square Gothic
rotesque	Olive X-Light	Square Gothic
rotesque	Olive X-Light	Square Gothic
rotesque	Olive Light	Square Gothic
rotesque	Olive Light	Square Gothic
rotesque	Olive Regular	Square Gothic
rry Thin	Olive Regular	Square Gothic
rry Heavy	Olive Medium	Square Gothic
rry Fat	Olive Medium	PEIGNOT LIGHT
vetica	Olive Bold	PEIGNOT LIGHT
elvetica	Olive Bold	Avant Garde Book Cond.
lvetica	Olive Black	Avant Garde X-Lite
lvetica	Olive Black	Avant Garde Book
lvetica	Olive Black	Avant Garde Med.
lvetica	Olive Black	Avant Garde Demi
lvetica	Olive Narrow	Avant Garde Bold
elvetica	Olive Narrow	Blippo Black
elvetica	Optima	Blippo Black Outline
elvetica	Optima	Soul Light
elvetica	Optima	Soul Medium
elvetica	Optima	Soul Bold

normal Typography

tight Typography

very tight Typography

touching Typography

normal Typography

tight Typography

very tight Typography

touching Typography

These two pages from Nortype's (Norwalk, Connecticut) type book show at a glance the kinds of spacing available these days with computer type.

This amazing (to any one old enough to remember when it wasn't available) development allows the designer to achieve a "look" that goes beyond the mere choice of type face. And since the same size can now be expanded or contracted, it allows you to deal with space problems in a wonderfully freer way.

Computer-Generated Foto-Typography

norType

Norwalk, Conn. 06856

Samples of

UNIT SPACING BETWEEN LETTERS

1. foto-typography — minus 2 units
2. foto-typography — minus 1 unit
3. foto-typography — set normal
4. foto-typography — plus 1 unit
5. foto-typography — plus 2 units

Foto-Typography offers a choice of options to produce optically pleasing interword and inter-character spacing. By the stroke of a key, spacing may be varied from tight to loose in increments of 1/32 of an em. When the ultimate in quality typography is required, letter combinations such as Te, Yo, LY can be set snugly in any size from 5 to 36 point. These samples demonstrate a few of the choices available to the artist or art director to custom-fit his typography.

set normal

Foto-Typography offers a choice of options to produce optically pleasing interword and inter-character spacing. By the stroke of a key, spacing may be varied from tight to loose in increments of 1/32 of an em. When the ultimate in quality typography is required, letter combinations such as Te, Yo, LY can be set snugly in any size from 5 to 36 point. These samples demonstrate a few of the choices available to the artist or art director to custom-fit his typography.

minus 1 unit

Foto-Typography offers a choice of options to produce optically pleasing inter-word and inter-character spacing. By the stroke of a key, spacing may be varied from tight to loose in increments of 1/32 of an em. When the ultimate in quality typography is required, letter combinations such as Te, Yo, LY can be set snugly in any size from 5 to 36 point. These samples demonstrate a few of the choices available to the artist or art director to custom-fit his typography.

plus 1 unit

Foto-Typography offers a choice of options to produce optically pleasing interword and inter-character spacing. By the stroke of a key, spacing may be varied from tight to loose in increments of 1/32 of an em. When the ultimate in quality typography is required, letter combinations such as Te, Yo, LY can be set snugly in any size from 5 to 36 point. These samples demonstrate a few of the choices available to the artist or art director to custom-fit his typography.

minus 2 units

Foto-Typography offers a choice of options to produce optically pleasing inter-word and inter-character spacing. By the stroke of a key, spacing may be varied from tight to loose in increments of 1/32 of an em. When the ultimate in quality typography is required, letter combinations such as Te, Yo, LY can be set snugly in any size from 5 to 36 point. These samples demonstrate a few of the choices available to the artist or art director to custom-fit his typography.

plus 2 units

ABOVE SAMPLES SET IN TWELVE POINT VEGA LIGHT—ONE POINT LEADED

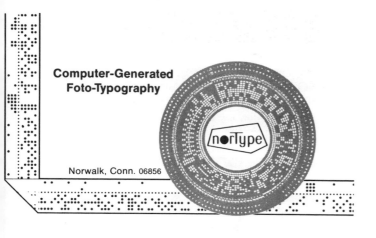

Calligraphia	T681
Calligraphia Outline	T681(O)
Central A-32	
Central A-51	
Central A-78	
Coronet Bold	
Kaufman Script	TK-2
Park Avenue	
Snell Roundhand	OS-113

G24	24pt.
G18	18pt.
G12	12pt.
G8	8pt.
G6	6pt.
G4	4pt.
G3	3pt.
G2	2pt.
G1	1pt.

These are the weights of rules that you will either be drawing by hand or doing with tape or ordering in type. Pt. stands for "point" and that's how you refer to it.

A B C D E F G H I J K
L M N O P Q R S T U
V W X Y Z & ! ? " " – ( ) *
abcdefghijklmnopqr stuvwx
xyz1234567890 $ $ ¢ %

These examples are from Franklin Photolettering, New York City

Torino Flare

AAAAA AABBCCDDEEEEFFGGHHH
LLLMMHHHHHHHIIJJJJKK KKKKK
MMMNNNNNNNOOPPQQRR RRRSS
SSSTTTTUUUUUVVVVVWWWWWXX
XXYYYYYZZ–!?&.,:;'""'*/()• aaa bbccccddd
eeeffffggghh hhii ijjkkk k lllmm
mnn nooppqqrrsss sttt tuuuvvwwwxxxyy
yzz % $$¢1234567890

Here are two examples of Swash alphabets—these are letters that are fancied up with curved flourishes. These faces usually give you a choice of certain letters so you can put together words wherein letters can tuck in or curve over or under.

SpaFlair

ABCDEEFGGHHIJJKKKLLLMMNNOPP
RRR SSTTTTUUVVVWWWXXYYYZZ(–!?&
.,:;'""/•) aaabccdddeeeefffgghhhiiiijj
kklllmmmmnnnnoooopp pqqrrrssttttttuu
uvvvvwwwwxxxyy yyyz%.,$¢1234567890

Other kinds of type and letter forms . . .

Newly designed transfer faces keep appearing and are available at art supply stores. They're expensive—but they're often mighty handy.

Many of the new faces are novelty ones—(often too complicated and tricky for extensive use)—but occasionally just the ticket for a short line or an initial cap or a logotype. For modest printed projects where you must keep the cost down you will find that typewriter type is satisfactory.

This is especially so if you have access to the electric typewriters—the kind with the ball head that give you several choices of typefaces.

The cost of this is considerably less than regular type. Its limitations are fewer choices of faces and many fewer exacting niceties. But when your funds are limited you do what you must do. It shows no lack of standards not to squander money on a piece that does not require it. If you design and prepare your piece well, the difference in type will go unnoticed by most recipients.

NEWSLETTER

September

Hope that........you all had a great summer........your tennis game improved.... your air conditioner worked.......the water wasn't polluted...........you didn't ge sunburned........none of your jobs bounced........and that you're now ready to tackle that $40,000 assignment in good humor.

As for us, since July we've been working on a $150,000 assignment in the basement. our new gallery! Plumbers are relocating all the steam, water and waste pipes, elec cians are rewiring for track lights, demolition crews are removing unneccessary wall a terrazzo floor is being poured, holes are being cut in the gallery floor for the stai wells to the lower level, the Permanent Collection storage room is being expanded an

---

These unusual faces are from the Mecanorma Collection, Keuffel & Esser Co.

FLEURDON

Black only

AB

61 6257-120C 120 Pt. Caps, Numerals

ABCD

61 6257-84C 84 Pt. Caps, Numerals

ABCDEF

61 6257-60C 60 Pt. Caps, Numerals

ABCDEFG

61 6257-48C 48 Pt. Caps, Numerals

FULL STROKE

Black only

AB

61 6239-84C 84 Pt. Caps, Numerals

ABC

61 6239-72C 72 Pt. Caps, Numerals

ABC

61 6239-60C 60 Pt. Caps, Numerals

NEWZELEK

Black only

ABCD

61 6242-84C 84 Pt. Caps, Numerals

ABCD

61 6242-72C 72 Pt. Caps, Numerals

ABCED

61 6242-60C 60 Pt. Caps, Numerals

PAPERCLIP CONTOUR

Black only

abcdefgh

61 6224-84L 84 Pt. LC, Numerals

abcdefghij

61 6224-72L 72 Pt. LC, Numerals

abcdefghijkl

61 6224-60L 60 Pt. LC, Numerals

abcdefghijklmno

61 6224-48L 48 Pt. LC, Numerals

There are three other methods of handling the printed word—all with limitations—all appropriate under the right conditions.

These are: Hand lettering, hand writing, or calligraphy, which is a kind of combination of both. See examples. Your judgement will tell you when it's right to use one or the other.

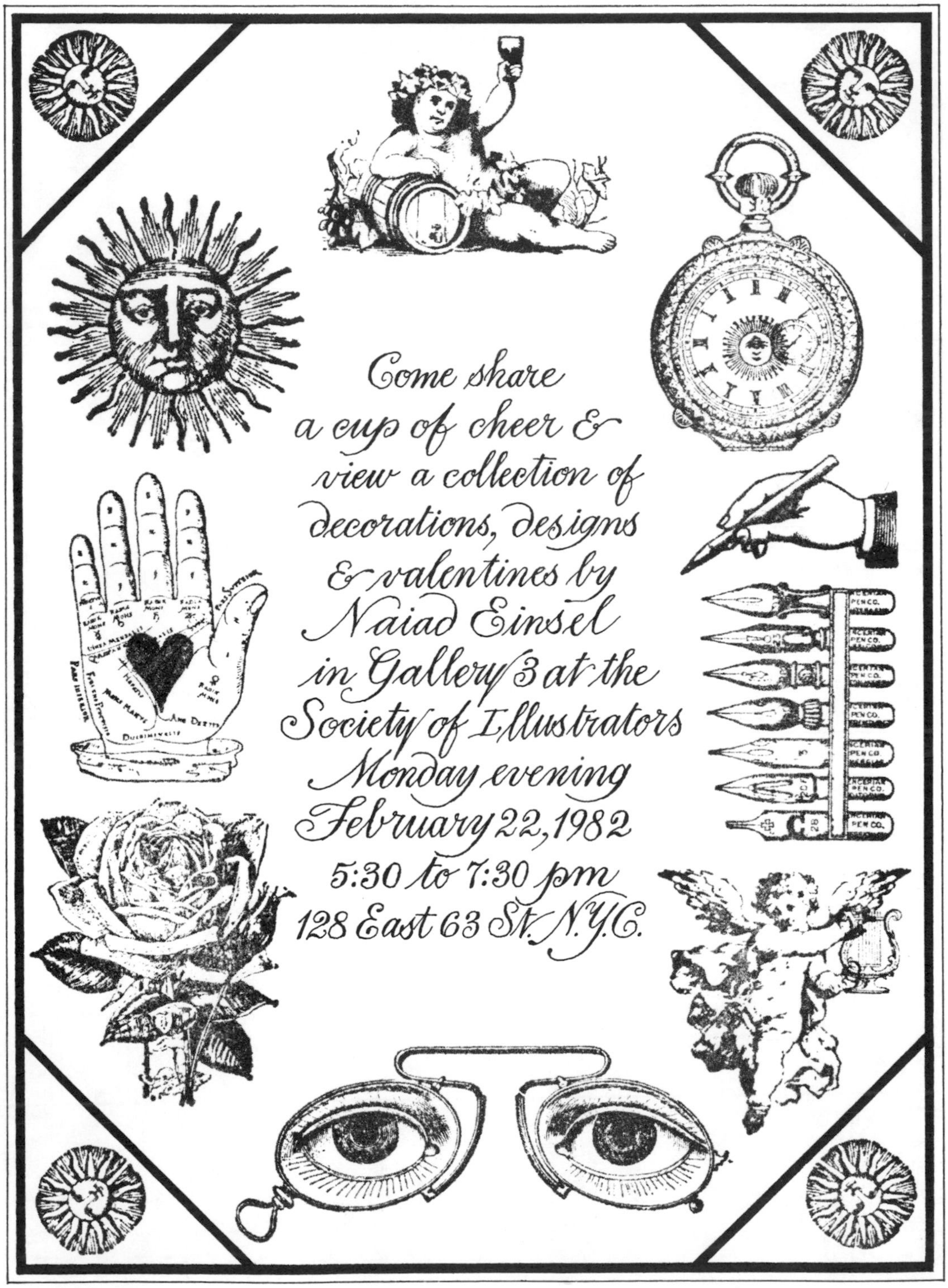

These calligraphy examples are the work of Bride Whelan who teaches it at the Paier College of Art in Hamden, Connecticut.

There is a great renewal of interest in calligraphy, and it's ironic that an art of the ancient monks is increasingly in demand for charts and presentations in today's corporate world. Ms. Whelan is prominent in this field.

I suggest you look into the study of this subject yourself. There are excellent books available and it is being taught all over the country. The new interest has also brought about many new types of pens and shaped markers to facilitate its practice.

You'll find it mighty handy to be able to do some of your copy by hand.

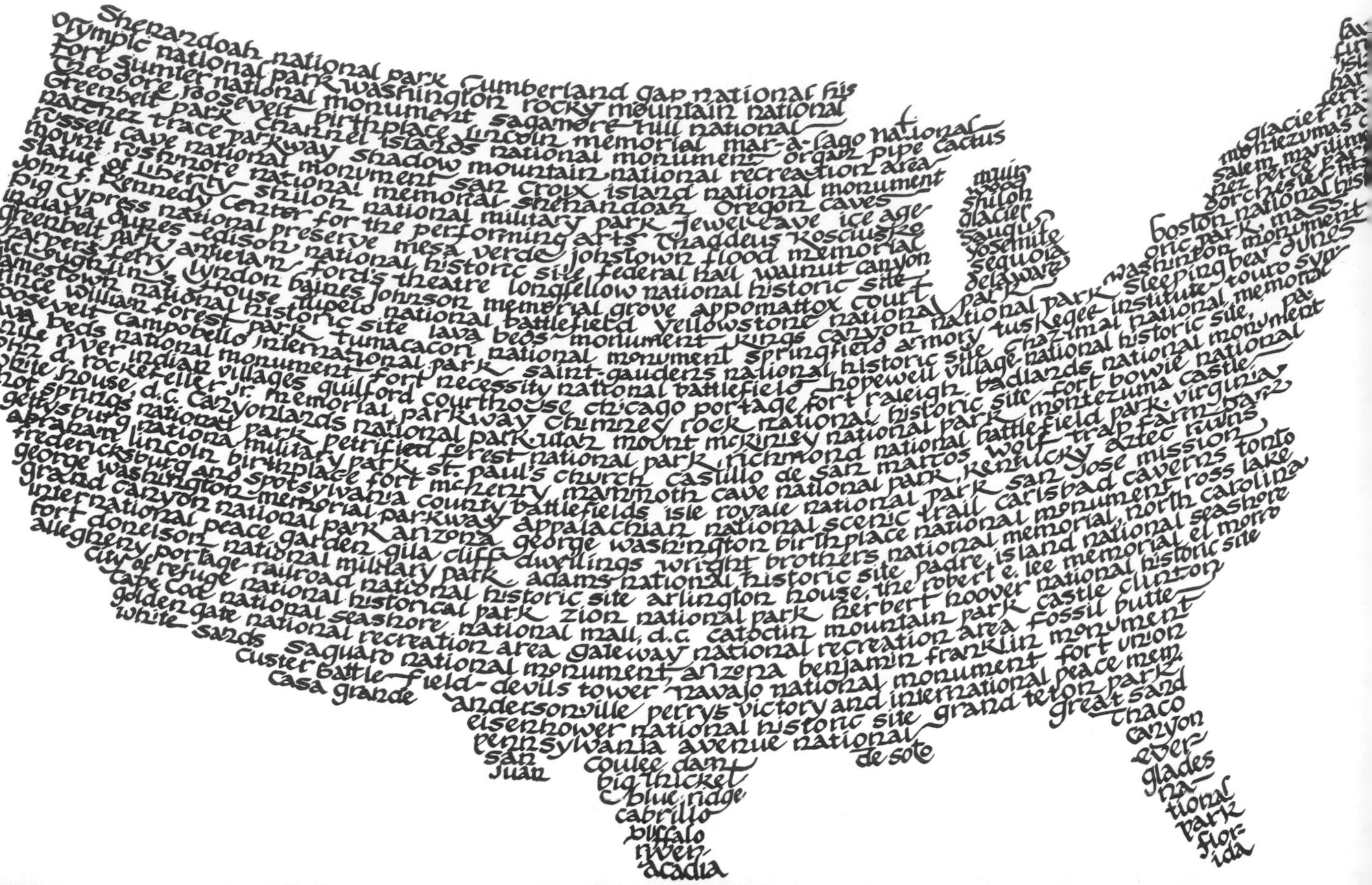
Shenandoah national park
olympic national park washington cumberland gap national his
fort sumter national monument rocky mountain national
Theodore Roosevelt birthplace sagamore hill national
Greenbelt park channel islands lincoln memorial mar-a-lago national
natchez trace parkway shadow mountain national monument organ pipe cactus
russell cave national monument san croix island national recreation area
statue of liberty shiloh national military park jewel cave ice age
John F. Kennedy Center for the performing arts Thaddeus Kosciuszko
big cypress national preserve mesa verde johnstown flood memorial
indiana dunes edison national historic site federal hall walnut canyon
muir
shiloh
glacier
saugus
yosemite
sequoia
delaware
glacier
boston
white sands saguaro national monument arizona benjamin franklin fort union
custer battlefield devils tower navajo national monument international peace park
casa grande
andersonville perrys victory and international peace mem.
eisenhower national historic site grand teton park great sand
pennsylvania avenue national
san juan
coulee dam
big thicket
blue ridge
cabrillo
buffalo river
acadia
de soto
chaco canyon
everglades national park florida

Observe the variation in tone—like a watercolor.

◀ *Note how different the same piece can look when you reverse it.*

TRICKS OF THE TRADE:
A few ingenious methods that allow you to arrive at findings you'll need to know.

CROPPING

There are two cropping devices available to you: one is manufactured and the other is homemade. Both work on the same principle. They each allow you to view your photo with as much off the top and bottom and both sides as you wish. You can adjust vertically and horizontally at the same time.

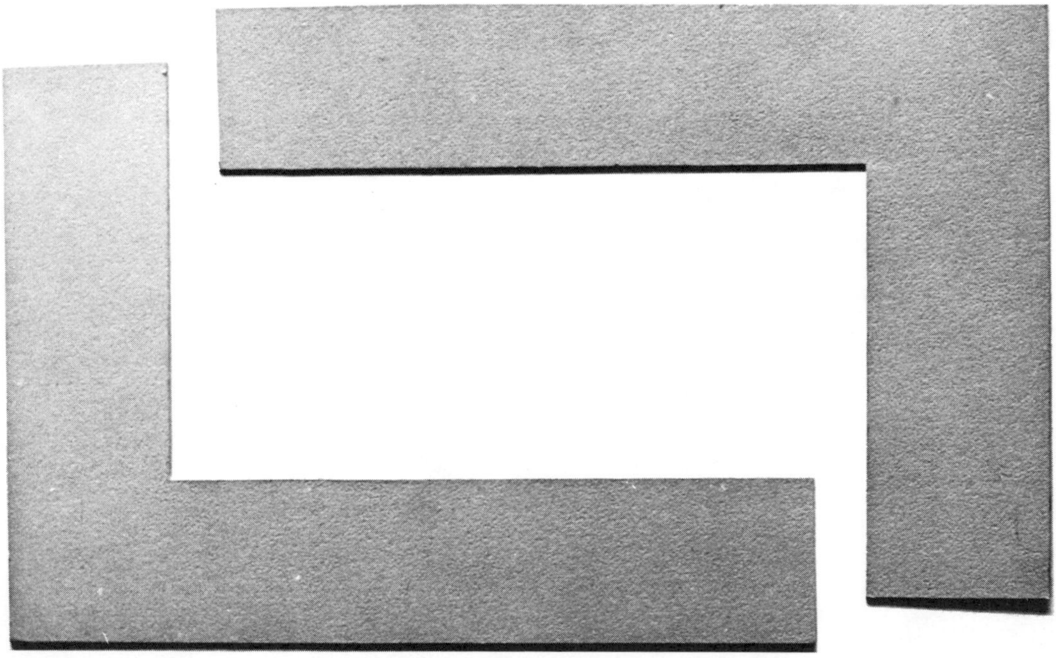

The handmade version consists of two L-shaped pieces of cardboard.

The full photo.

Lay them both on the photo and work the left one "down" and the other "up." Move them back and forth until you're satisfied. Then place a piece of tracing paper over the opening and gently trace the image and the four corners. Always take care not to press down on photos with pencils and not to mark on photos with anything but a grease pencil, which will later rub off.

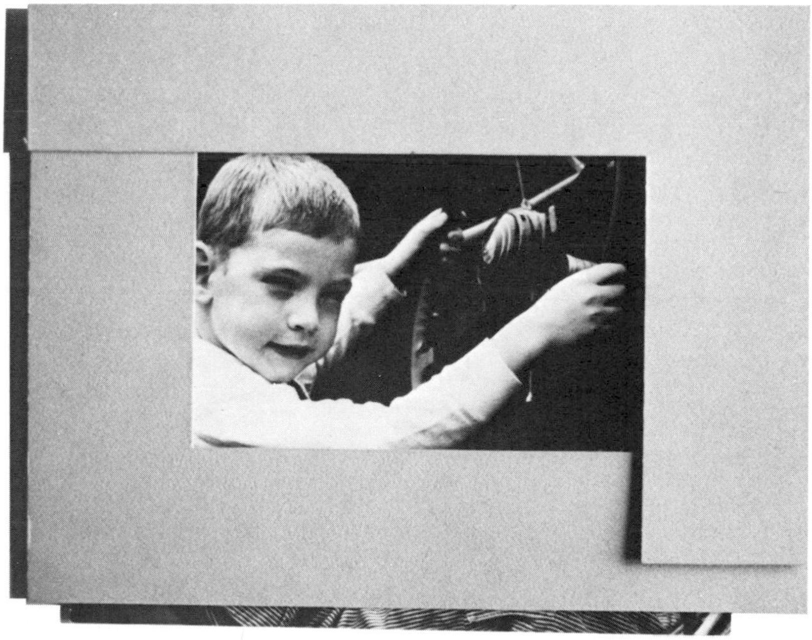

The desired cropping.

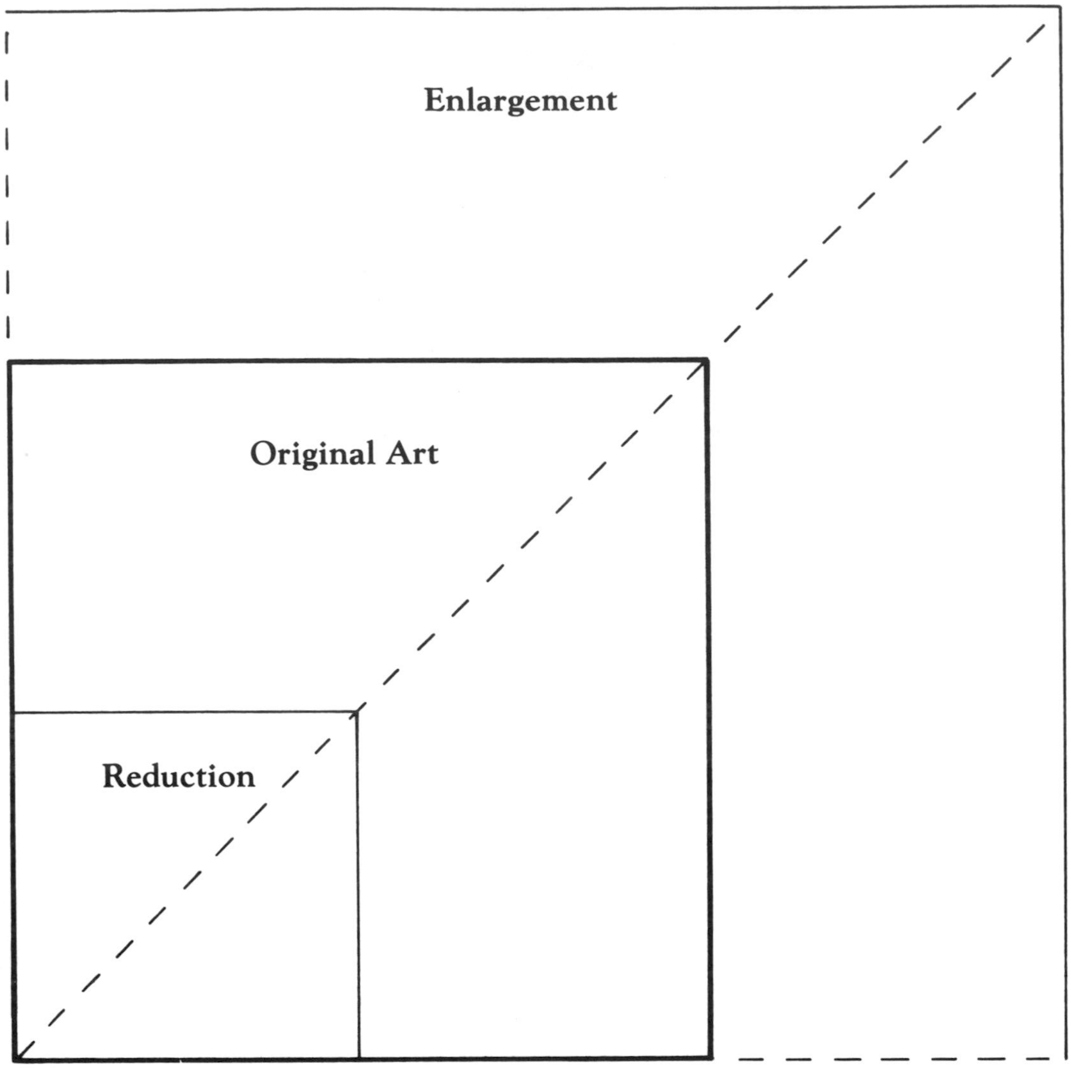

How to scale pictures up or down in size.

It seldom happens that the photo or art work you're dealing with is the size it will need to be when reproduced.

Scaling it up or down is very simple: place a piece of tracing paper over the picture. Then rule a line from the lower left corner through the upper right corner. If you wish to enlarge, extend the line until it reaches the height you wish the picture to be. Now drop a vertical line down to the bottom line and a horizontal line over to the left, that's your new scale. To reduce in scale, follow the same procedure inside *the original size of the picture.*

If you need to scale a silhouette, draw a rectangle around its top, bottom and sides with a blue pencil and proceed as above.

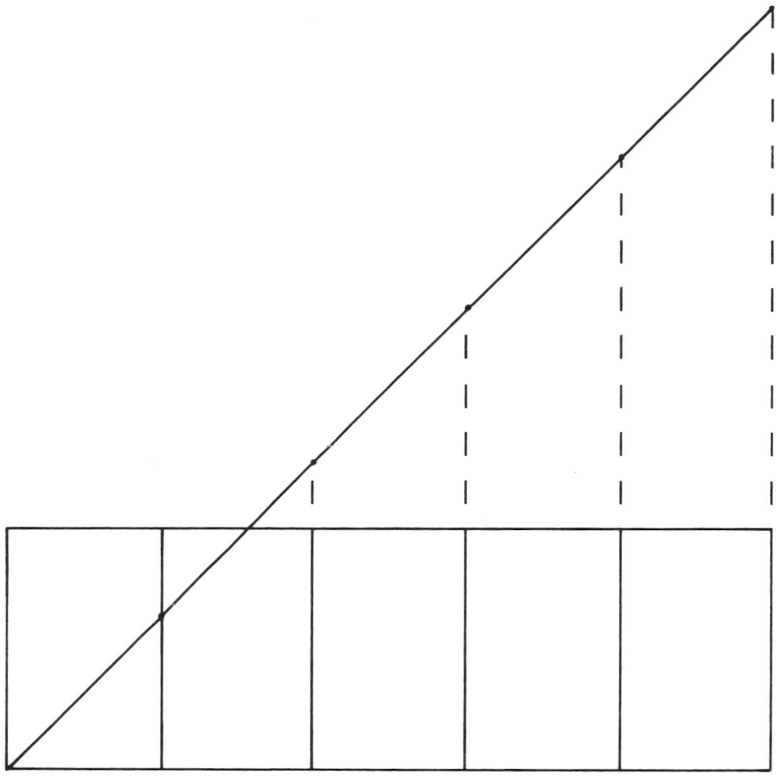

Dividing areas into even spaces.

This is a method of dividing uneven areas into even parts. To demonstrate, we'll deal with a rectangle 7 inches wide and 2¼ inches deep (shown reduced).

Assume we wish to divide it into five identical spaces vertically.

First extend the last line up. Then place your ruler from the lower left corner until it touches the 10-inch mark (you choose 10 because 5 is divisable into it). Now tick off every 2 inches and drop vertical lines down.

To determine horizontal spaces extend the top *line to the* right *and proceed as described above.*

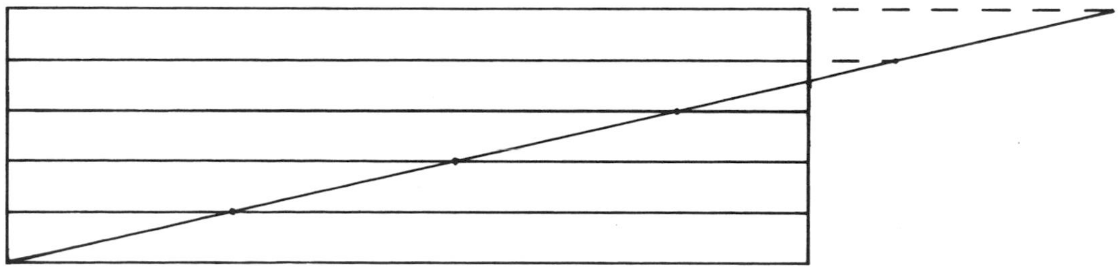

COPY FITTING

The very first requirement is neatly typed copy. Don't have type set from handwritten material, no matter how legible it may be.

There are many methods of calculating the area that your typewritten copy will occupy when it's set in type.

However different the approaches might be, they all depend on *one* and *the same thing*: that is to know how many typograghic *characters* you're dealing with.

A *character* is any of the following:

an individual letter
a punctuation mark
a space between words or punctuation marks

When you count characters, those are the *only* things to count.

Let's try it.

Put a sheet of typewritten copy before you—even a typed letter will do.

1 Lightly draw a line down the right side through the average length lines. (We'll come back to the leftover words to the right of your lines later.)

2 Count the number of characters and spaces in an average single line (up to your pencil line).

3 Count the number of lines on the page, and add a line or two to account for the leftovers to the right of your pencil line.

4 Now multiply the characters per line by the number of lines on the page. The answer gives you the number of characters to a page. (If you had several pages typed the same length you'd then multiply a single page's finding by the number of pages.)

5 Using the pica measure on your Haberule, measure the width of the copy block that you've indicated on your layout. The width in picas will now be known as the "measure."

6 Open your type book to the typeface and the size you'd like to use. Most type books have a table that quickly shows you the number of characters in various pica widths. If you're obtaining your first type book, try to get one with such a table. If you don't have the table, merely mark off the desired pica width on the example in your type book, then count those characters. It will give you the same answer that a table would.

7 When you know the number of type-set characters to your pica width, divide that number into the multiplied number above. That will tell you how many lines deep your copy will come.

8 Using the Haberule again, select the vertical slot next to the type size you chose. Now read down to the figure you've just learned by what you did in No. 7. That will show you how deep your copy will come.

It's as simple as that—except for one other small thing: unless you specify otherwise, type will be set solid. That means with as little space between lines as mechanically possible. You can have as much space as you want by adding what's called leading. It's specified by the point.

Sometimes you use lead for cosmetic reasons—other times to flesh out copy when you have space to spare.

9 If you were to add 1 point of lead to a 10 pt. typeface you would calculate your depth on the Haberule as 11 point—and so on up if you use more lead.

10 pt. type 2 pt. lead - 12 pt. on Haberule
10 pt. type 3 pt. lead - 13 pt. on Haberule
10 pt. type 4 pt. lead - 14 pt. on Haberule
(or 7 pt. x 2)

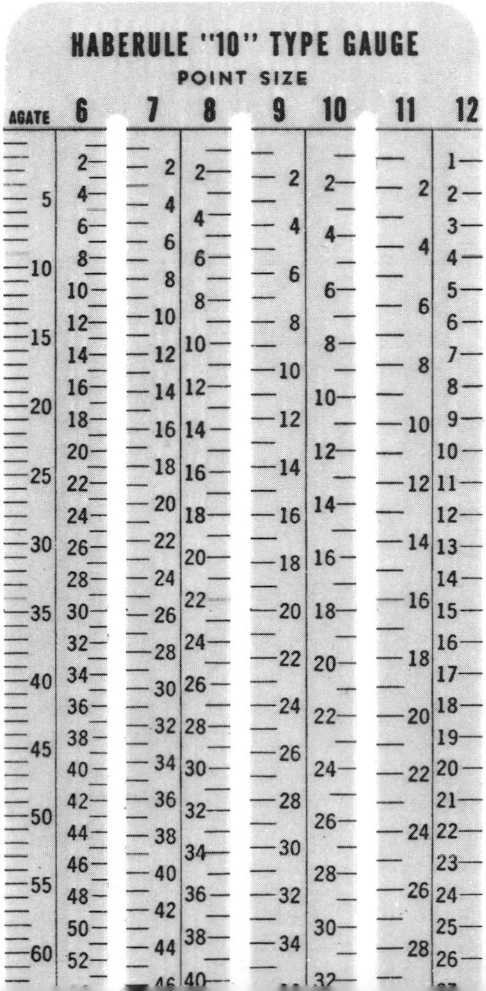

Type Size

12 pt. edge is also a pica rule.

Read down the appropriate slot to learn how many lines deep your copy will come.

Shown reduced

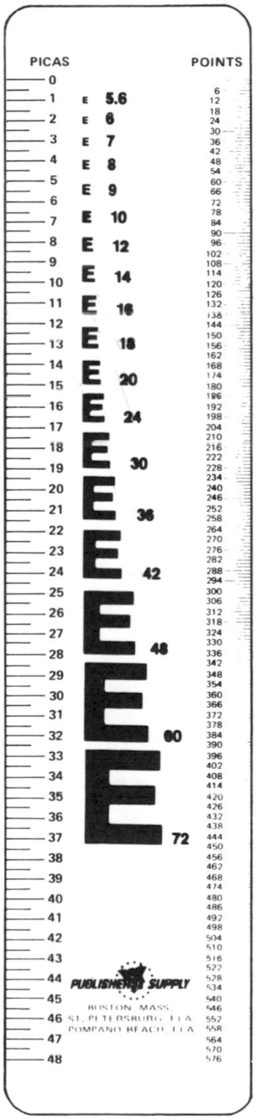

This clear device lets you discover the size of a type you may not know. You simply lay one of the E's over an E in the type in question. When it matches exactly, it is the point size shown at the bottom right of the E.

12/13 Goudy ital. X 33 pi Fl. L. only

no indent

8 pt # space

One of the nice treats designers always have awaiting them is the opening paragraph of a column of text or copy. It can be the icing on the cake -- the cherry on the sundae.

As you can see by the several examples shown here, there is an endless variety of visual ways to begin your type -- it's up to you and your inventiveness.

Among other ways, you can begin with a bold version of the typeface you're using (if there is one) or you can begin with a bold or decorative initial of any style of your choosing.

It can rise above the first line or dip as deeply below it as you wish.

The first full thought can be set in caps. You can begin with a black spot called a "bullet"; these can be square or round -- solid or outline.

You can even use a decorative drawing that contains the first letter of the first word in your copy.

Your type file should contain lots of decorative letters to employ for opening lines.

Don't allow opportunities to slip by -- it's wise to always have a good opening act.

This is a typical piece of copy marked up and ready for setting.

The strange looking hieroglyphics are really very simple: they name the typeface. They give its size and the amount of leading. They give the width (measure) and tell whether the copy is to be set flush left, flush right, or flush left and right.

The instructions also indicate the paragraph indentations.

This is the copy for the captions on the right-hand page: Examine the two and see how the type boils down from the double-spaced typewriter version.

ON Olin Court in the new town of Jonathan, Minn., there stand a few houses that may be the building blocks of new architectural trends in residential housing for years to come.

These prototype houses, product of a special study by Stanford Research Institute and development by Jonathan Housing Corporation, are modular flexible dwellings with a

They say there are more sheep than people in New Zealand, and an examination of the statistics quickly shows this to be a gross understatement. While there are about three million people in those Pacific islands, there are 60-million sheep, as well as 8,500,000 cattle. No other country has nearly so much livestock in proportion to human beings.

Some of Samuel Rubin's friends jokingly suggest that he go out of the bread business and concentrate on selling posters. Mr. Rubin's advertising campaign, "You don't have to be Jewish to love Levy's real Jewish rye," has won customers not only for his New York-baked bread but also for his ad posters. They are selling in bookstores, card shops and department

■ Once I was afraid of auctions. My first, at a city gallery, made me feel poor—people were bidding more on vases and paintings than I had earned in five years of hard work. My second one was in a

As fresh as a windswept forest, pine furniture is enjoying new popularity—and in styles so warm and sophisticated that you'll be taking a new look at the wood that was the mainstay of the American home more than 200 years ago. Today, pine is more than

IN THE FALL OF 1961 I walked into the faded-brown lobby of my Zurich hotel, the Savoy, alive with a sense of adventure. Within minutes I was speaking on the telephone with Topic Mimara himself. I was incredibly

The rainy season can be a treacherous time to travel in Panama. I learned that lesson rather suddenly while driving in the lush, green highlands of Chiriquí, Panama's westernmost province. The small Datsun station wagon bounced over the roughly paved, two-lane highway that runs north from the

HIRING, promotion, demotion, and transfer decisions are among the most difficult and important that managers are required to make. Not only do these actions affect the success and even the survival of the organization; they also involve the careers of individuals.

In addition, managers will find that their decisions in these matters frequently must be approved by superiors, and that peers

All the children were silent now. The three youngest stood in a tight corner near the sink, their eyes grave, their mouths open, saliva glistening on their lips.

One of the nice treats designers always have awaiting them is the opening paragraph of a column of text or copy. It can be the icing on the cake the cherry on the sundae.

As you can see by the several examples shown here, there is an endless variety of visual ways to begin your type—it's up to you and your inventiveness.

Among other ways, you can begin with a bold version of the typeface you're using (if there is one) or you can begin with a bold or decorative initial of any style of your choosing.

It can rise above the first line or dip as deeply below it as you wish.

The first full thought can be set in caps. You can begin with a black spot called a "bullet"; it can be square or round—solid or outline.

You can even use a decorative drawing that contains the first letter of the first word in your copy.

Your type file should contain lots of decorative letters to employ for opening lines.

Don't allow opportunities to slip by—it's wise to always have a good opening act.

THE MECHANICAL:

**Otherwise known as the paste-up.
This is the carpentry of the graphic arts.
Herein several simple practice exercises to
train your hands to perform for love &/or money.**

The almost universal use of the offset printing process has led to a demand for an endless supply of "Mechanical Men"—more accurately, the Mechanical Woman, since females now predominate in this field. Gender is unimportant. What is important is that the person doing the mechanicals be able to work neatly and accurately.

A short exposure to the tools, the procedures and purpose of a mechanical, plus practice in cutting, pasting, ruling and other bits of graphic carpentry will hone your skills surprisingly fast.

The function of the mechanical is to accurately assemble all of the elements of a job onto one manageable piece or pieces of white cardboard, and to include instructions in the printer's language so that he may proceed without your presence. This is called "Camera-Ready Art."

In addition to positioning elements, the mechanical person must be able to do repair work on type; such things as correcting typos, improving spacing, re-positioning individual letters or whole lines of type.

This is all done with either a single-edged razor blade, or an X-acto knife plus a tweezer. Good mechanical people can deftly pick out and put back tiny specks of punctuation—a period from here—a comma from there—and presto!—a newly born semicolon out of two things the size of a gnat's eye!

I think this would be a good time for you to familiarize yourself with your materials and to begin to practice those skills you'll need to acquire.

First, you'll need the proper "paper" to work on. In this case it will be illustration board, which is actually drawing paper glued to cardboard for rigidity.

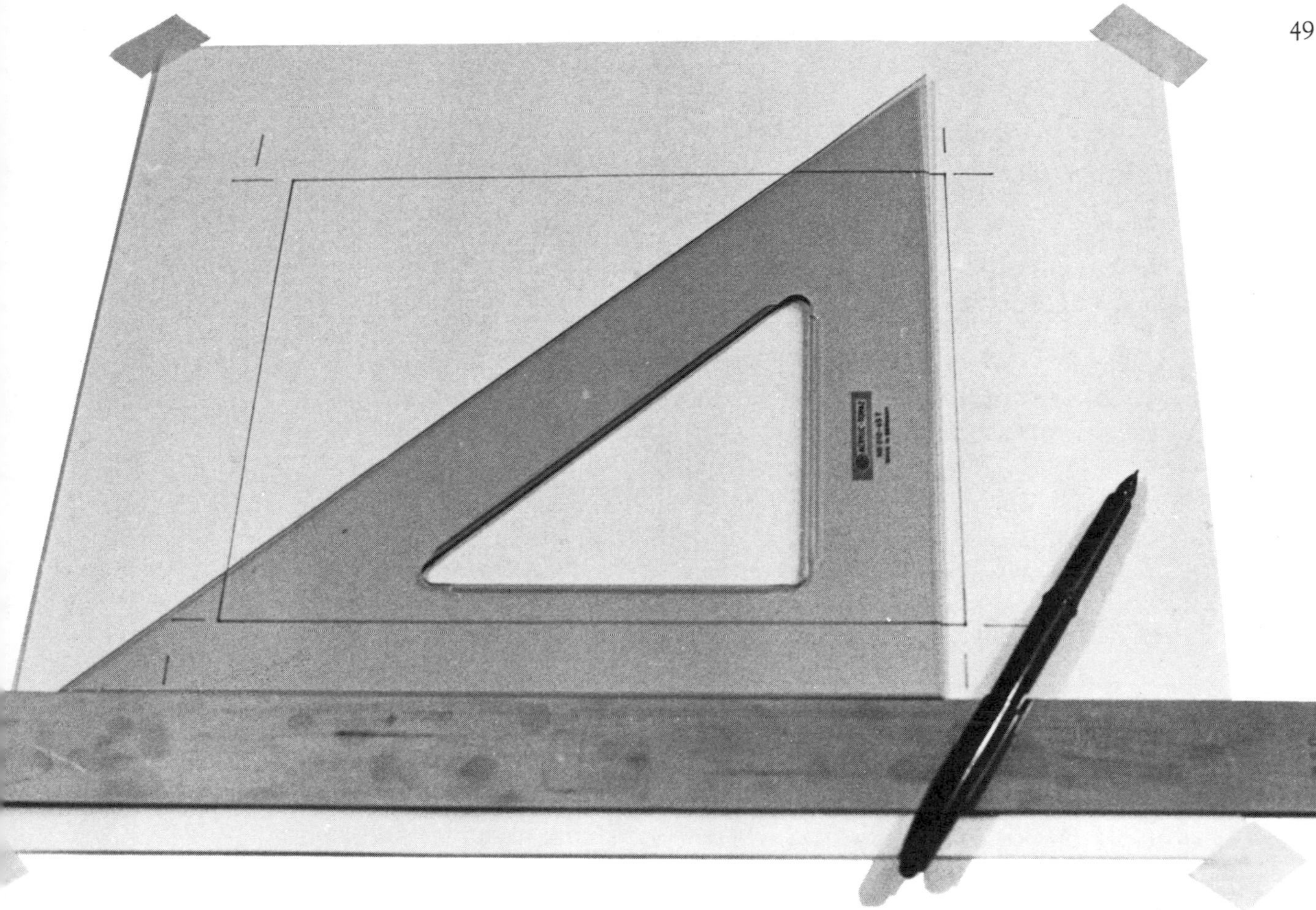

There are many brands of board of equal worth. Two that many professionals use are Crescent #200 and Bainbridge #172. Your art supply store may carry other kinds. Ask for illustration or mechanical board, single weight, *hot pressed.* That means smooth, as opposed to cold pressed which has a textured surface. Trial and error and price will help you decide your favorite.

I want you now to go through some exercises all aimed at the things you need to be able to perform in doing a mechanical.

Set out the following tools and materials: your T-square, a medium to hard blue pencil, a triangle, a steel ruler, a ruling pen (or a set of mechanical pens), a pencil compass and an ink compass, and a single-edged razor or X-acto knife.

A word of explanation about the blue pencil: faint blue is not picked up by the printer's camera; consequently (and happily), you can use it for guide lines and not have to erase. Just be sure it's well-sharpened and that you don't bear down any heavier than you need to. You'll be ruling with ink right on top of it; and since the line is wax, it will foul your pen and spoil your rule if it's any heavier than a ghost mark.

BEGIN THIS WAY:

(1) Clear the decks of all things that might impede the freedom of movement of your hands or your instruments. Make certain, also, that the surface *under* your work is flat and smooth—free from deep cut marks, paper clips and such. Get in the habit of changing your protective cardboard as soon as it begins to get scarred; a bumpy surface will affect your ruling.

(2) Line up the top or bottom of your mechanical board horizontally with your T-square.

(3) Fasten it firmly with a 1-inch piece of masking tape on all four sides.

(4) Mark off 8 inches horizontally, with two light dots of blue. Now drop the T-square down about 10 inches, and with your triangle, lightly draw two connecting vertical lines. They'll be the boundary lines that you will rule from and to.

Think of the look of a ruled note pad where the lines are about ½-inch apart. That's what you're about to practice.

Now, fill your ruling pen. Adjust it for a very fine line.

Remember your T-square and triangle should have a strip of masking tape on the under side to keep the ruling edge slightly above the drawing surface. Starting bravely, and with a smooth stroke against your T-square, rule lines to the other side. Two don'ts. *Don't* change the angle of your wrist as you travel, and *don't* change your speed. It's not fatal if you pass by the mark because you can later touch it up with white. But it is calamitous if you fall short, because it's almost impossible to rejoin the line without a variation.

If you have a set of mechanical pens, proceed the same way, remembering to hold the pen upright. Experiment with the different size pens to learn the weights of line they produce.

Practice the above until you can consistently rule the kind of line you *want* —not the kind you'll settle for.

Here it gets a bit tougher; you must now rule the vertical lines to meet the horizontals—*exactly.* With your triangle set firmly against your T-square, rule from corner to corner.

Everybody finds their own style in best doing this. I like to stand up at the side of my drawing board so it's still possible to rule from left to right. Experiment until you can do it your way.

Try some inked circles now. Do them in pencil first. Decide on a size, and spread your pencil compass to *half* that size. Set the pin into the paper just firmly enough so it won't jump the track—but not so deep that it will leave an unsightly hole. Next, load your ink compass, adjust, and swing it in a circle. Practice all sizes and all weights. Your mishaps, such as ink flooding out—or giving out—or ends not meeting perfectly will reveal what needs correcting and more practice.

Templates for doing circles with mechanical pens are available, but they're limited to relatively small sizes. However, you should have one on hand.

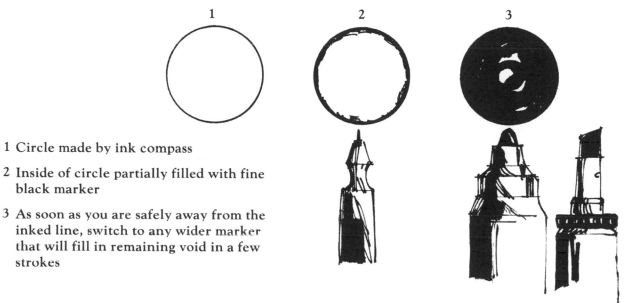

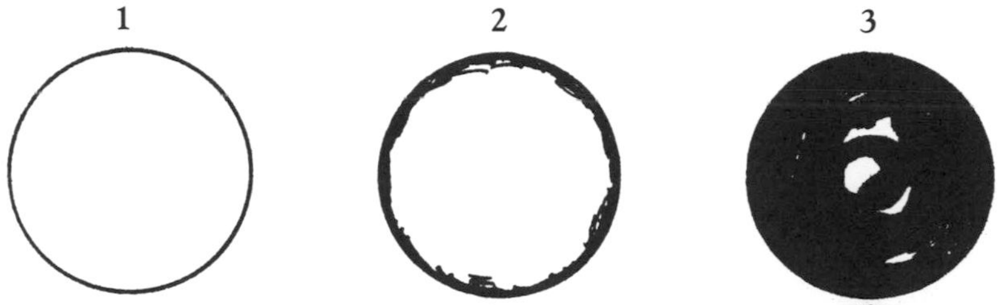

1 **Circle made by ink compass**

2 **Inside of circle partially filled with fine black marker**

3 **As soon as you are safely away from the inked line, switch to any wider marker that will fill in remaining void in a few strokes**

Next, rule in some ink squares about 3 inches all around. Now, try filling them in solid black. Do the same with some of the circles you made.

Here is a fine instance of what a boon the felt marker in all its variations has been to the commercial art world; it's obvious that the important thing to avoid in filling in the squares and circles is not to go outside the inked line. So, with a relatively fine marker, carefully fill in about $\frac{1}{16}$ inch from the line inwards. When you're safely within the line, select a broader marker and work towards the center.

Always keep at least three widths of oil-based black markers on hand. The reason for using these as opposed to water-based markers is that the former can be touched up with white without bleeding through.

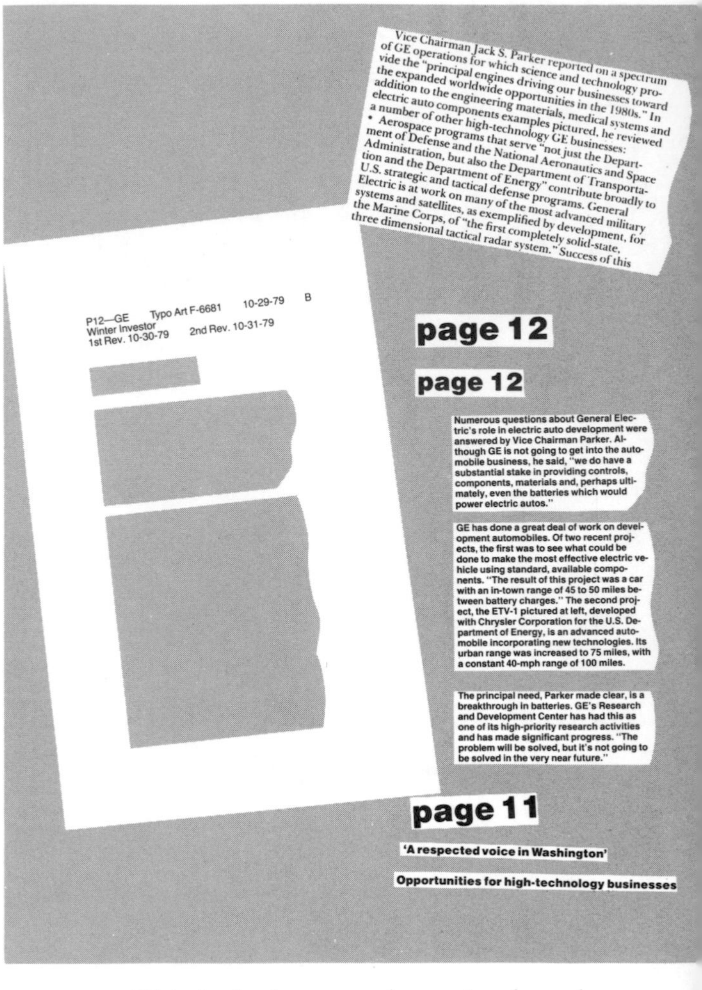

This is the homemade cutting board.

All the pieces on the board have one coat of rubber cement on the back. They stick just enough so they won't move as they're cut.

These four pieces have been cut from the proof on the left (now it can be discarded).

Everything is ready to be picked up with your blade and transferred into position on your mechanical board.

When you set type down "by eye" be sure not to press it firmly until it has been squared. When the page is complete and everything has been checked, place a piece of clean paper over everything — then with the broad side of your triangle, wipe it firmly from edge to edge. This is known as burnishing — its purpose: to fix everything in place with no loose edges or wobbly pieces to get out of square.

This seems an appropriate place to introduce you to a helpful studio trick that someone once showed me; it's a homemade board for cutting out type.

Start with a heavy piece of cardboard about 9 x 12 inches (the chip board used as backs for paper pads are perfect). Cover it with strips of wide masking tape, being careful to butt each succeeding strip snugly against the other with no overlap. The tape should be longer than the board so it will wrap around the edges. The secret of this rig is that the tape has a waxy residue, and when you put a *dry* rubber cement surface against it, it holds firmly, but it also lifts off easily after you've made your cuts. It works like a gummed label.

At this stage it's unlikely that you'll have access to real type proofs to practice with, so do the next best thing; take pages from magazines or junk mail or old greeting cards. Use anything except newspapers, just so long as it contains lines of type. Put a coat of cement on the back, let it dry, and *lightly* place it on your cutting board. Then, line your T-square up below a line of copy *as close as you can get without touching the type*. Now cut. Repeat this over and under several lines of type until you become proficient at it. When you've cut out several lines, lift the excess paper and discard it. What remains is ready for you to reach for when the time comes to place it on your mechanical.

Here's further practice: rule another ink rectangle 4 x 6 inches. Draw a blue line down the center—then find the center of each line of type. Now, pick up a line of type with your razor blade and place the center mark down lightly to match the center line; then, using the T-square, line it up and press it down firmly when it's square. Repeat this until you've filled the space to your satisfaction.

Now practice this: go back and juxtapose some individual letters, being sure to space and square them properly.

When you've mastered ruling and manipulating type you will have learned the manual basics of doing a mechanical.

Almost everything else will be procedure.

---

page 12 page 12

This is an example of correcting space via your eye and your blade.

Note how the left version is full of "air" holes— whereas the corrected one flows uninterruptedly.

a world premiere

SATURDAY EVENING OCTOBER 8 • 8:30 at the

HARTMAN THEATRE COMPANY
61 ATLANTIC STREET
STAMFORD, CONNECTICUT

STRIP IN PHOTO "A"

Reverse this area

THE 1977-78 SEASON REPETOIRE:

CHARADE
Music: Carl Maria von Weber
Choreography: Genia Melikova

LES SYLPHIDES
Music: Chopin
Choreography: Michel Fokine

MYTH
Music: Stravinsky
Choreography: Alvin Ailey

GRAND PAS de DEUX CLASSIQUE
Music: Aubee
Choreography: Victor Gaovsky

SPRING WATER
Music: Rachmaninoff
Choreography: Asaf Messerer

DON QUIXOTE
Music: Minkus
Choreography: Marius Petipa

PAS DE QUATRE
Music: Pugni
Choreography: Perrot

ROSE ADAGIO
Music: Tchaikovsky
Choreography: Marius Petipa

DESIGN FOR SIX
Music: Tchaikovsky
Choreography: John Taras

BLACK SWAN PAS DE DEUX
Music: Tchaikovsky
Choreography: Petipa

GENIA MELIKOVA Artistic Director

THE BERNHARD BALLET TICKET FORM
Saturday, October 8th 8:30 CURTAIN

Single ticket prices at $8.00, $6.00. A $2.00 discount for Children, Students, Senior Citizens.
HARTMAN SUBSCRIBERS: A 13% discount off ticket prices.

PLEASE RESERVE FOR ME:

......tickets at $8.00 or......13% discount ....$________
......tickets at $6.00 or......13% discount ....$________
......tickets for Children, Student, Seniors ....$________
TOTAL $________

NAME ....STREET.....................

CITY .................STATE..........ZIP..........DAYTIME PHONE..........

......I AM A HARTMAN SUBSCRIBER.

......I enclose my check made payable to The Bernhard Ballet. A non-profit professional organization.

Please send check to Hartman Theater Co., 61 Atlantic Street, Stamford, Conn. 06901.
BOX OFFICE PHONE: (203) 323-2131. Box Office will be open September 26th 10 a.m. to 5 p.m. Mondays thru Saturdays. Please enclose a stamped, self-addressed envelope.

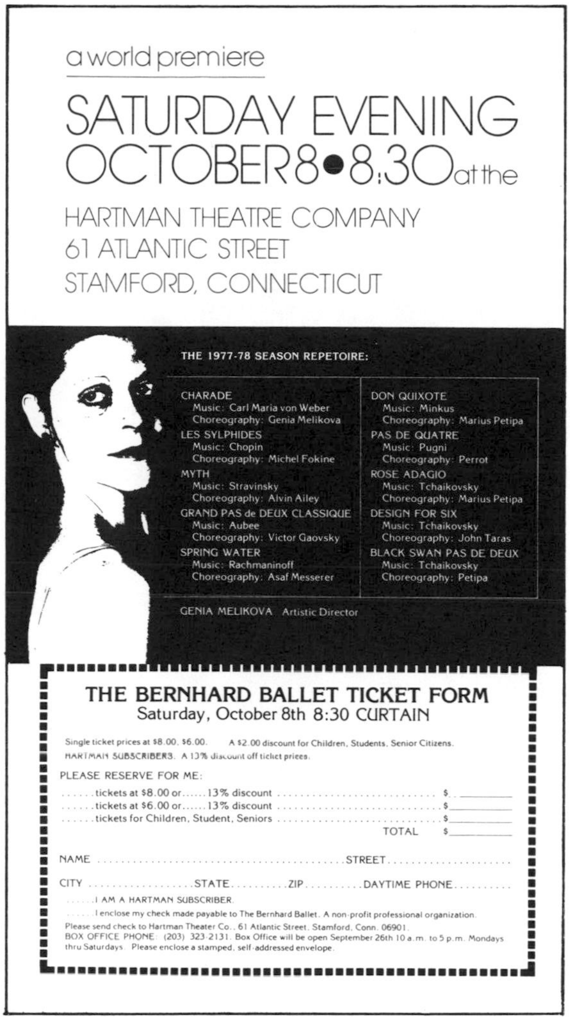

a world premiere

SATURDAY EVENING
OCTOBER 8 • 8:30 at the

HARTMAN THEATRE COMPANY
61 ATLANTIC STREET
STAMFORD, CONNECTICUT

THE 1977-78 SEASON REPETOIRE:

CHARADE
Music: Carl Maria von Weber
Choreography: Genia Melikova

LES SYLPHIDES
Music: Chopin
Choreography: Michel Fokine

MYTH
Music: Stravinsky
Choreography: Alvin Ailey

GRAND PAS de DEUX CLASSIQUE
Music: Aubee
Choreography: Victor Gaovsky

SPRING WATER
Music: Rachmaninoff
Choreography: Asaf Messerer

DON QUIXOTE
Music: Minkus
Choreography: Marius Petipa

PAS DE QUATRE
Music: Pugni
Choreography: Perrot

ROSE ADAGIO
Music: Tchaikovsky
Choreography: Marius Petipa

DESIGN FOR SIX
Music: Tchaikovsky
Choreography: John Taras

BLACK SWAN PAS DE DEUX
Music: Tchaikovsky
Choreography: Petipa

GENIA MELIKOVA Artistic Director

THE BERNHARD BALLET TICKET FORM
Saturday, October 8th 8:30 CURTAIN

Single ticket prices at $8.00, $6.00. A $2.00 discount for Children, Students, Senior Citizens.
HARTMAN SUBSCRIBERS. A 13% discount off ticket prices.

PLEASE RESERVE FOR ME:

......tickets at $8.00 or......13% discount .... $______
......tickets at $6.00 or......13% discount .... $______
......tickets for Children, Student, Seniors .... $______
TOTAL $______

NAME .... STREET ....

CITY .... STATE .... ZIP .... DAYTIME PHONE ....

......I AM A HARTMAN SUBSCRIBER.

......I enclose my check made payable to The Bernhard Ballet. A non-profit professional organization.

Please send check to Hartman Theater Co., 61 Atlantic Street, Stamford, Conn. 06901.
BOX OFFICE PHONE: (203) 323-2131. Box Office will be open September 26th 10 a.m. to 5 p.m. Mondays thru Saturdays. Please enclose a stamped, self-addressed envelope.

Reproduced version.

Here, and on the next page, are two "camera-ready" mechanicals, which means "ready for the engraver's camera."

Everything is where it should be. The size is correct. The bleed marks at the corners are in place. The fold lines have been indicated by a broken rule. It has been checked against the dummy to see that it will fold into the proper places. (Sometimes pages must be placed upside-down so they'll fall into place when folded.)

Although this job was printed in line, it has been reproduced here in halftone to allow you to see the cut-marks and the space adjustments.

Note how the center panel has been reversed. That instruction and all others are noted in the margins.

MAINE ARTISTS

For its seventh mid-summer exhibition, the General Electric Gallery presents the work of five distinguished artists who live, work and teach at least a part of the year in the State of Maine. They are not, however, limited to regional perspectives.

If fact, the subject matter of these extensively-traveled painters is universal. The works of Messrs. Betts, Etnier and Thon come to us courtesy of the Midtown Galleries in New York City. Mr. Chadbourn's are lent by the Jacques Seligmann Galleries, N.Y., Shore Gallery, Boston, Mass., and the Barridoff Gallery, Portland, Maine. Mr. Muench's prints come courtesy of Associated American Artists, N.Y.

May
July
1980

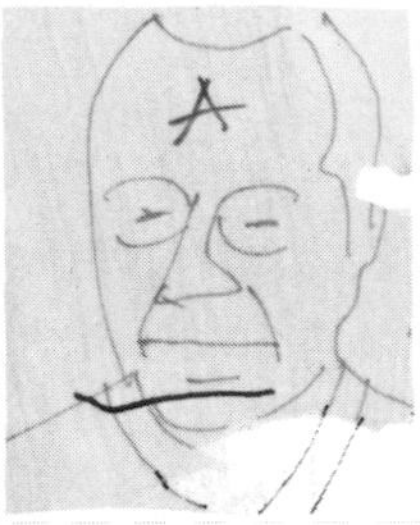

EDWARD BETTS
of Ogunquit

Edward Betts was born in 1920 at Yonkers, New York. He received his B.A. at Yale University in 1942 and his M.F.A. at the University of Illinois in 1952. He also studied at the Art Students League in New York. The artist lives in Illinois during the school year and at Ogunquit, during the summer.

Over the years many honors have come to Betts, a highly respected Professor of Art at the University of Illinois. He has won numerous top awards in national painting and watercolor exhibitions and is represented in many major museum and private collections. His abstractions based on Maine coast subject matter, in lacquer, collage, mixed media, and watercolor have been seen in many one-man shows, and national exhibitions. Of his one-man show at Midtown Galleries, one critic wrote, "...lacquers and oils, drawings and gouaches and collages, all employing the most exquisite and brilliant colors. For the most part, the paintings of Mr. Betts appear to be skillfully controlled explosions of angular forms whose colors produce an exceptionally exciting effect. Not to be overlooked are the collages, which have a most unusual richness and strength."

ALFRED C. CHADBOURN
of Yarmouth

Born in Izmir, Turkey, of American parents, he studied at the Chouinard Arts Institute, Los Angeles, Academie De La Grande Chaumiere and Ecole Des Beaux Arts, Paris.

He has had a score of One-Man Shows in America and Europe and has been the recipient of numerous awards.

His works are in the private collections of Mr. and Mrs. Douglas Dillon, Jean Cocteau, Sir Lawrence Olivier, Jacques Cousteau, Princess Grace of Monaco, Walter Cronkite and others. Among corporate collections containing his works are those of Philip Morris, Inc., McGraw-Hill, and Connecticut Bank and Trust Company.

His teaching experience includes associations with the Pasadena Art Institute, Queen's College, and Portland School of Art.

Elected to the National Academy in 1970, he is represented by the Jacques Seligmann Galleries, New York City, Shore Gallery, Boston, and the Barridoff Gallery, Portland, Maine.

STEPHEN ETNIER
of South Hampswell

Born in York, Pennsylvania, he attended Yale and Haverford College briefly before studying at the Pennsylvania Academy of Fine Arts. Later studies included work with Rockwell Kent and John Carroll. His wartime service included duty with the U.S. Naval Reserve as commanding officer of several convoy vessels.

His works are included in collections of the Metropolitan Museum, Los Angeles Museum, Toledo Museum, Boston Museum, Yale University Art Gallery, Vassar College Art Gallery, Greenwich Art Museum and others.

Among his many awards are the Samuel F.B. Morse Gold Medal, the Benjamin Altman Prize, Saltus Gold Medal — all at the National Academy of Design — and the Butler Art Institute Purchase Prize.

He holds Doctor of Fine Arts Degrees from Bates and Bowdoin Colleges in Maine.

JOHN MUENCH
of Freeport

A native of Medford, Mass., he studied at the Academie Julien, Paris, the Art Students League and with Mourlot and Dorfinant in Paris.

His wide teaching experience includes associations with The Contemporaries Graphic Arts Workshop, NYC; Portland, Maine, School of Art; Impressions Workshop, Boston; Rhode Island School of Design; Boston University; and Westbrook College. Currently he is Artist in Residence at Westbrook College in Portland and with the Maine Printmaking Workshop.

He has been granted seven fellowships and has been included in over 200 group shows in the United States, Europe, Japan and New East as well as numerous AFA traveling exhibitions and art for U.S. embassies. His 32 one-man shows include those at the Smithsonian Institution and National Collection of Fine Arts.

Muench has been honored with 24 awards for painting, drawing and prints.

WILLIAM THON
of Port Clyde

Born in New York City, he is self-taught. He holds an Honorary Doctor of Arts Degree from Bates College, was elected Trustee of the American Academy in Rome (1966) and elected a member of the National Institute of Arts and Letters (1967).

He has taught and lectured at the Atlanta Art Association, John Herron Art Institute, Indianapolis, Ind., Munson-Williams-Proctor Institute, Utica, NY, Norton Gallery and School of Art, W. Palm Beach, Fla., and was artist in residence at the American Academy, Rome.

His work is included in a variety of collections, including the prestigious Joseph H. Hirshhorn Collection and those of William Benton, Robert F. Woolworth, General Lucius Clay, Garry Moore, Dinah Shore, Mrs. Laurence S. Rockefeller. Other collections include those of the Metropolitan Museum, Whitney Museum, Pennsylvania Academy, Farnsworth Museum, Swope Gallery, National Academy of Design, Parke Davis & Co., Irving Trust Co., First National City Bank, Salomon Brothers & Hutzler and International Business Machines.

The lesson to be learned here is how to key photos to their position on the mechanical. It's simple; put the same letter on the back of the photo as you put in its allotted space.

Note that neither of these photos appear in the mechanical or in the printed piece as they are in the original. Each has been cropped.

Observe that Chadbourn's head has been enlarged—Muench's has been reduced.

Final note: the line indications for size on the mechanical were done on an apparatus called a "Lucey," which enlarges and reduces. The same result could be accomplished with photostats and marked "For position only."

V

⁄IAINE ARTISTS

or its seventh mid-summer xhibition, the General Elec- ric Gallery presents the work f five distinguished artists ʼho live, work and teach at east a part of the year in the tate of Maine. They are not, owever, limited to regional erspectives.

fact, the subject matter of ese extensively-traveled ainters is universal. The orks of Messrs. Betts, Etnier nd Thon come to us courtesy f the Midtown Galleries in ew York City. Mr. Chad- ourn's are lent by the Jacques eligmann Galleries, N.Y., hore Gallery, Boston, Mass., nd the Barridoff Gallery, ortland, Maine. Mr. Muench's rints come courtesy of ssociated American Artists, .Y.C.

May
July
1980

EDWARD BETTS
of Ogunquit

Edward Betts was born in 1920 at Yonkers, New York. He received his B.A. at Yale University in 1942 and his M.F.A. at the University of Illinois in 1952. He also studied at the Art Students League in New York. The artist lives in Illinois during the school year and at Ogunquit, during the summer.

Over the years many honors have come to Betts, a highly respected Professor of Art at the University of Illinois. He has won numerous top awards in national painting and watercolor exhibitions and is represented in many major museum and private collections. His abstractions based on Maine coast subject matter, in lacquer, collage, mixed media, and watercolor have been seen in many one-man shows, and national exhibitions. Of his one-man show at Midtown Galleries, one critic wrote, "...lacquers and oils, drawings and gouaches and collages, all employing the most exquisite and brilliant colors. For the most part, the paintings of Mr. Betts appear to be skillfully controlled explosions of angular forms whose colors produce an exceptionally exciting effect. Not to be overlooked are the collages, which have a most unusual richness and strength."

ALFRED C. CHADBOURN
of Yarmouth

Born in Izmir, Turkey, of American parents, he studied at the Chouinard Arts Institute, Los Angeles, Academie De La Grande Chaumiere and Ecole Des Beaux Arts, Paris.

He has had a score of One-Man Shows in America and Europe and has been the recipient of numerous awards.

His works are in the private collections of Mr. and Mrs. Douglas Dillon, Jean Cocteau, Sir Lawrence Olivier, Jacques Cousteau, Princess Grace of Monaco, Walter Cronkite and others. Among corporate collections containing his works are those of Philip Morris, Inc., McGraw-Hill, and Connecticut Bank and Trust Company.

His teaching experience includes associations with the Pasadena Art Institute, Queen's College, and Portland School of Art.

Elected to the National Academy in 1970, he is represented by the Jacques Seligmann Galleries, New York City, Shore Gallery, Boston, and the Barridoff Gallery, Portland, Maine.

STEPHEN ETNIER
of South Hampswell

Born in York, Pennsylvania, he attended Yale and Haverford College briefly before studying at the Pennsylvania Academy of Fine Arts. Later studies included work with Rockwell Kent and John Carroll. His wartime service included duty with the U.S. Naval Reserve as commanding officer of several convoy vessels.

His works are included in collections of the Metropolitan Museum, Los Angeles Museum, Toledo Museum, Boston Museum, Yale University Art Gallery, Vassar College Art Gallery, Greenwich Art Museum and others.

Among his many awards are the Samuel F.B. Morse Gold Medal, the Benjamin Altman Prize, Saltus Gold Medal — all at the National Academy of Design and the Butler Art Institute Purchase Prize.

He holds Doctor of Fine Arts Degrees from Bates and Bowdoin Colleges in Maine.

JOHN MUENCH
of Freeport

A native of Medford, Mass., he studied at the Academie Julien, Paris, the Art Students League and with Mourlot and Dorfinant in Paris.

His wide teaching experience includes associations with The Contemporaries Graphic Arts Workshop, NYC; Portland, Maine, School of Art; Impressions Workshop, Boston; Rhode Island School of Design; Boston University; and Westbrook College. Currently he is Artist in Residence at Westbrook College in Portland and with the Maine Printmaking Workshop.

He has been granted seven fellowships and has been included in over 200 group shows in the United States, Europe, Japan and New East as well as numerous AFA traveling exhibitions and art for U.S. embassies. His 32 one-man shows include those at the Smithsonian Institution and National Collection of Fine Arts.

Muench has been honored with 24 awards for painting, drawing and prints.

WILLIAM THON
of Port Clyde

Born in New York City, he is self-taught. He holds an Honorary Doctor of Arts Degree from Bates College, was elected Trustee of the American Academy in Rome (1966) and elected a member of the National Institute of Arts and Letters (1967).

He has taught and lectured at the Atlanta Art Association, John Herron Art Institute, Indianapolis, Ind., Munson-Williams-Proctor Institute, Utica, NY, Norton Gallery and School of Art, W. Palm Beach, Fla., and was artist in residence at the American Academy, Rome.

His work is included in a variety of collections, including the prestigious Joseph H. Hirshhorn Collection and those of William Benton, Robert F. Woolworth, General Lucius Clay, Garry Moore, Dinah Shore, Mrs. Laurence S. Rockefeller. Other collections include those of the Metropolitan Museum, Whitney Museum, Pennsylvania Academy, Farnsworth Museum, Swope Gallery, National Academy of Design, Parke Davis & Co., Irving Trust Co., First National City Bank, Salomon Brothers & Hutzler and International Business Machines.

THE DUMMY:

The above word in this case is neither a bridge term nor an I.Q. score. It's graphic jargon. Read on.

The word Dummy refers to a handmade facsimile of the printed piece you intend to produce.

It has two functions: one is to allow your client to see what the finished product will eventually look like; its secondary purpose is to allow the printer to estimate the overall price, and then to use it as a guide for size, folds, color, pages and numbers of photos and art. It should contain all answers to everything to do with the design and printing of the piece.

Dummys are often done roughly. This is o.k. if they're done for knowledgeable, experienced professionals who don't need things prettified—they only need facts.

On the other hand, some dummys, especially annual reports, are dressed up to such a degree that they appear to be finished. All it takes is enough money to hire expert hands.

For our purposes, however, we shall only concern ourselves with the homely, practical version. We just want to get on with the job with no wasted showmanship.

Two are shown here: the one on top is an educated scribble done to size. It shows the positioning of the color photos that are to occupy those areas. There is just enough drawing in them to identify which picture is which. It also shows the folds and the sequence of pages.

This was shown to a knowledgeable client who didn't require it to be rendered any more explicitly than it was.

The other dummy, done by Roman Mayer of Mayer Martin Studios, New York City, goes a large step farther. Though it is breezily rendered and the lettering roughly laid in, one can immediately envision its final look—and go straight to work producing it. For it is so deftly done that the style of cartoon is clearly suggested—the weight and style of type is apparent—the color breakup is shown—the folds and form are to actual size—the amount of body copy is precisely indicated, as is the area for the photos.

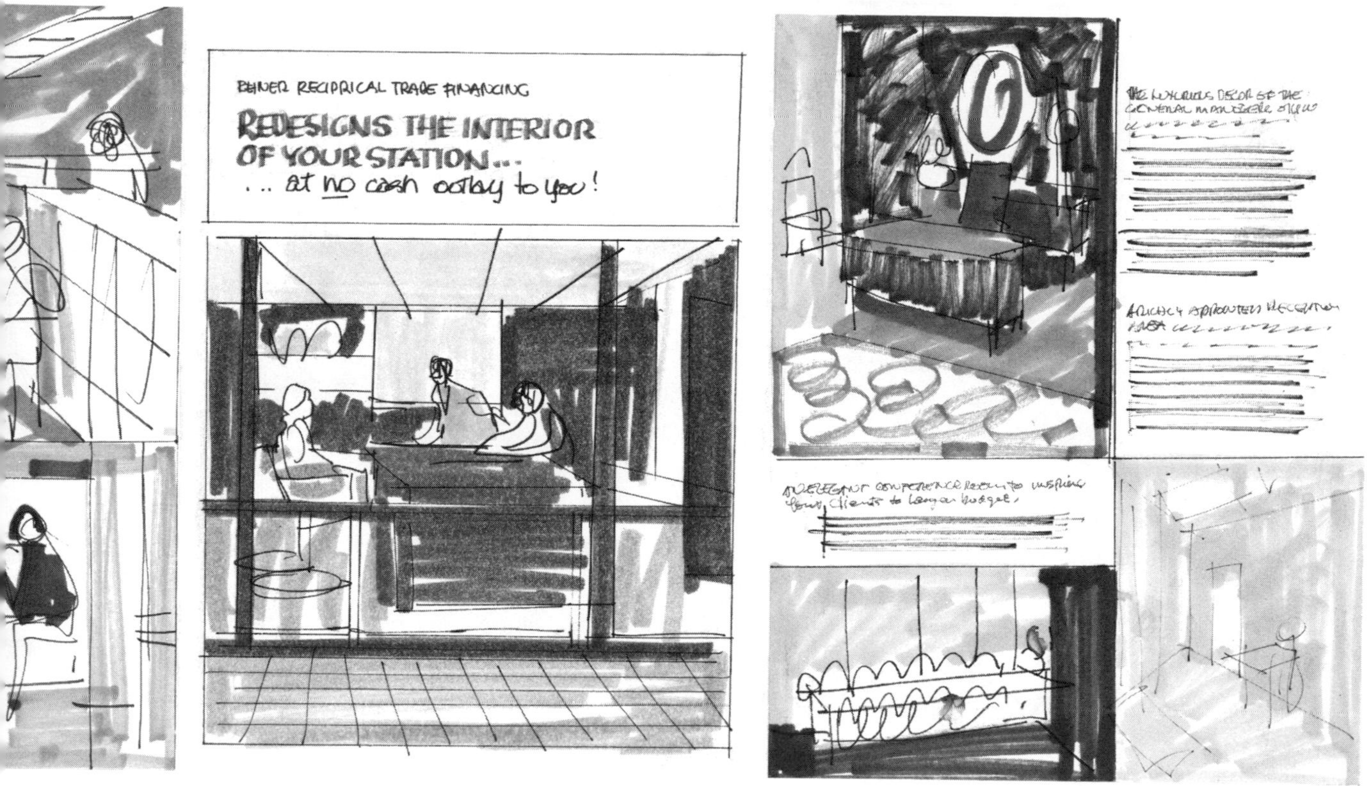

THE FRONT COVER

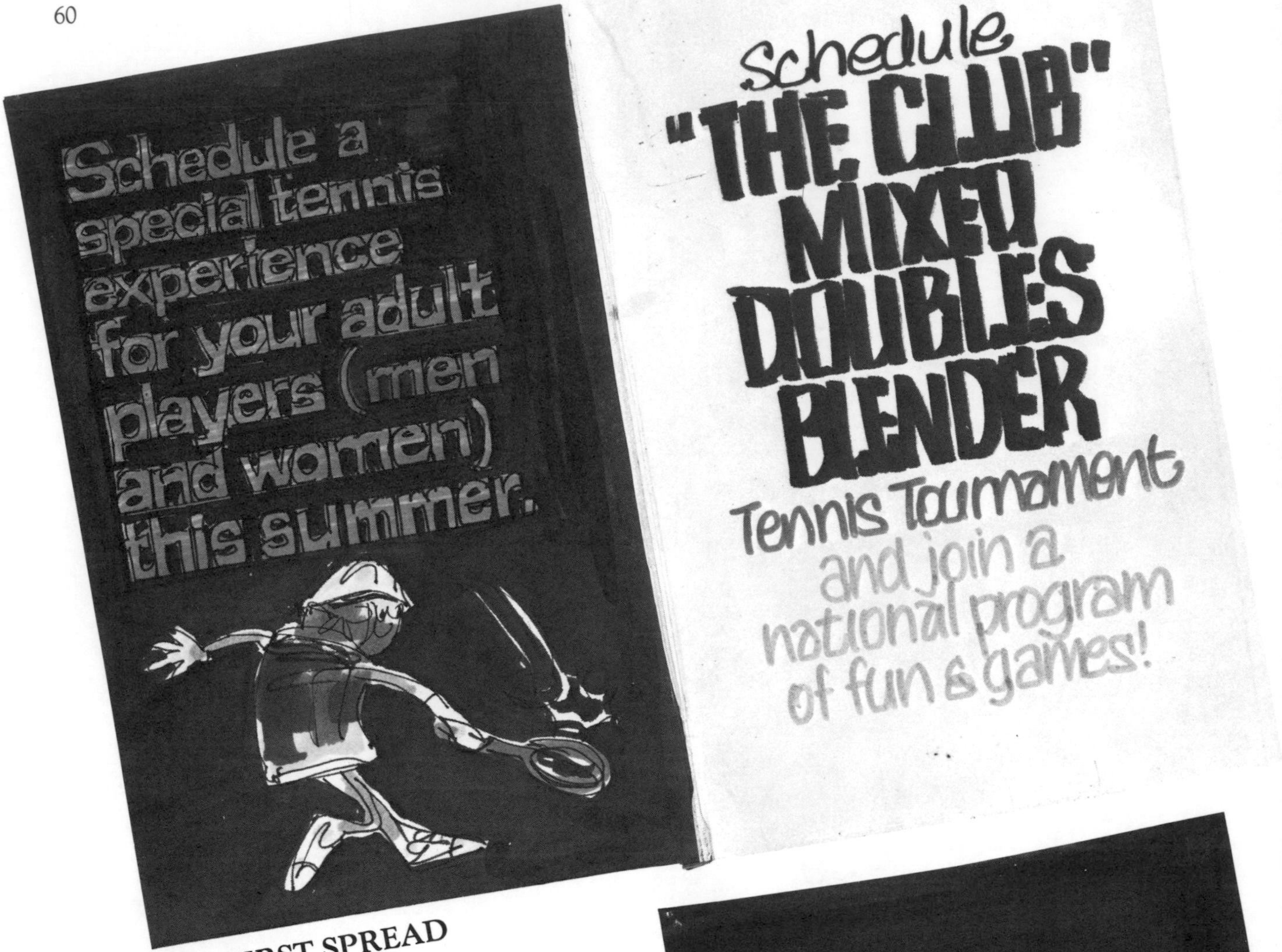

Schedule a special tennis experience for your adult players (men and women) this summer.
Schedule
"THE CLUB"
MIXED DOUBLES BLENDER
Tennis Tournament
and join a national program of fun & games!

THE FIRST SPREAD

THE BACK COVER

THE CENTER, UNFOLDED

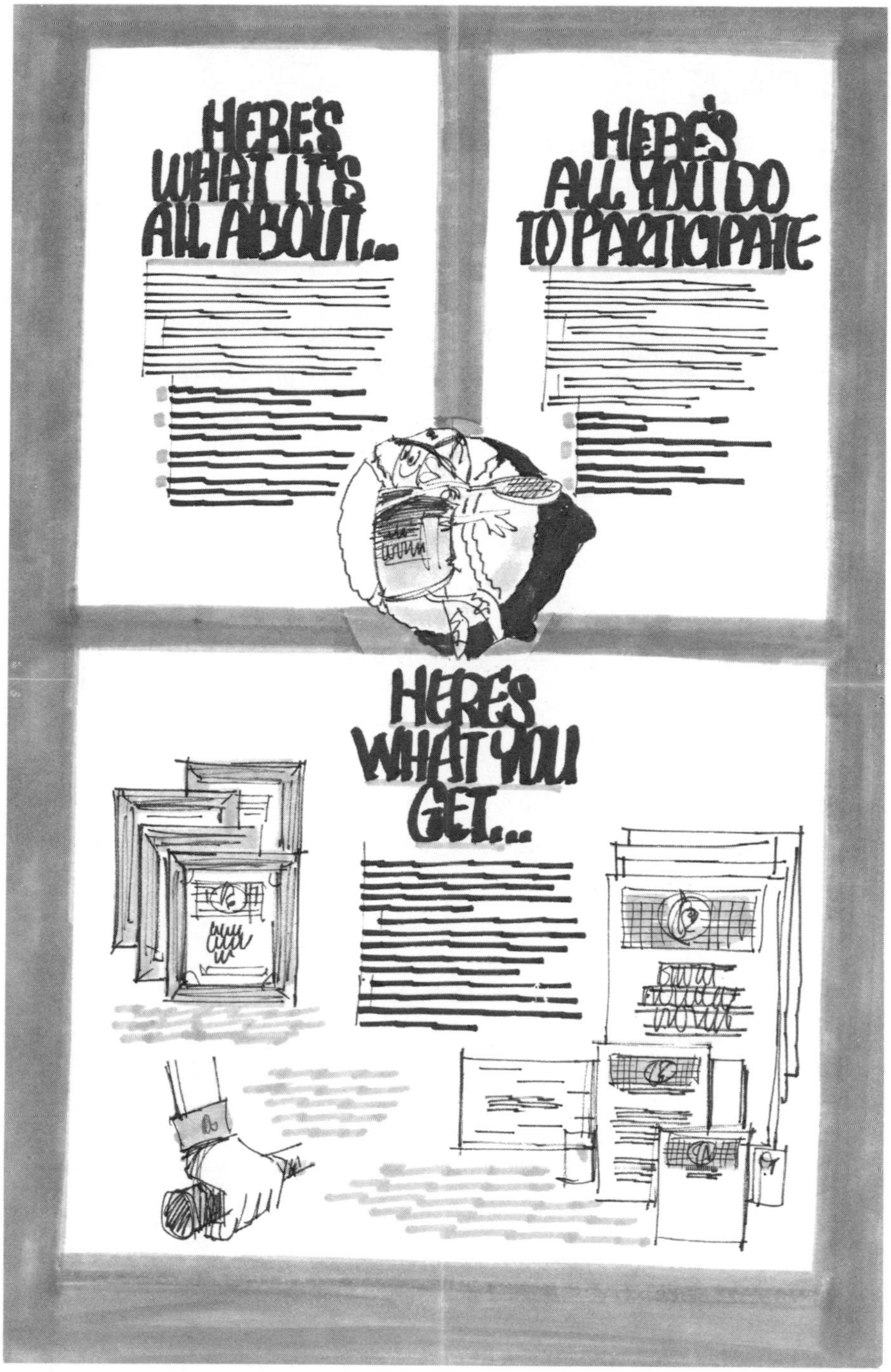

Don't be discouraged by the sureness of this professional job, which is probably well beyond your reach. But you too can fashion a usable dummy by working actual size and folding the piece in the logical sequence you want it to read. Letter the copy as well as you're able—making the letters as high as shown in your type book and as wide as the number of letters in a line. From this you'll be able to order the size of the display type you wish—and you'll be sure it will fit.

PRINTING PAPER:

A look into the delights & pitfalls of choosing the right & affordable paper for your assignments.

If it's true that "clothes make the man," as the old cliché states (now updated to read "clothes make the person"), then it can also be said that "paper makes the job."

The analogy doesn't bear too much scrutiny, of course, since many other factors enter into every printed piece. But it is true that paper is the initial cosmetic sight, color and touch we react to. The paper's texture—smooth, ribbed or glossy, produces an immediate esthetic sensation—good or bad.

It is, alas, also expensive, *very* expensive, out of sight or unavailable. Sometimes the paper you've selected from a book of samples has been discontinued. Often you can't have it because you don't need enough to make the costly shipping worth the manufacturer's while. Most often they won't split cases for small jobs.

But if you deal with a printer of any size, he will always have a varied supply of leftover papers for you to choose from. At times you can pick up great bargains and happy surprises this way.

You can also select several different papers and split the run. Apropos of this, I have discovered an unexpected reward: if your printed piece is to be displayed in stacks as pick-up items, they take on an added attraction when several colors are arrayed together. I do this when I have bookmarks printed for the public library. They look like fresh confections when they're set out on the counter along side each other. You can do the same thing with mailing pieces—especially self-mailers where you don't have the problem of matching envelopes.

A word of caution about paper weight; current postal charges are steep—and climbing. So beware of the beguiling heavy stock that *feels* so good and will *cost* so much if you exceed the weight limit.

Until you've gained experience, be sure to check with your printer about three other important aspects of paper: one is to design your piece at a size that leaves as little waste from a full sheet as possible—second, design it to fit standard sized envelopes. Custom-made ones, naturally, are expensive. And finally, design your job to fit as small a size press as is practical. For instance, a piece 11 x 17 inches will print beautifully on the small Multilith press. Its only limitation within that size is that it cannot handle jobs with large solid areas of color. The coverage is apt to be poor. In such cases, you must step up to a larger and more expensive press.

Paper manufacturers are very promotion conscious and most of them have programs for sending samples to designers. You can keep abreast of new developments in paper design by reading the ads in such trade magazines as: *PRINT*, *Graphics*, *Design USA*, *Art Direction* and *North Light*. Many of these are available in libraries. If your library doesn't carry them, check with your printer who is usually inundated with samples from paper houses. As you acquire paper samples—make them a part of your general graphics file.

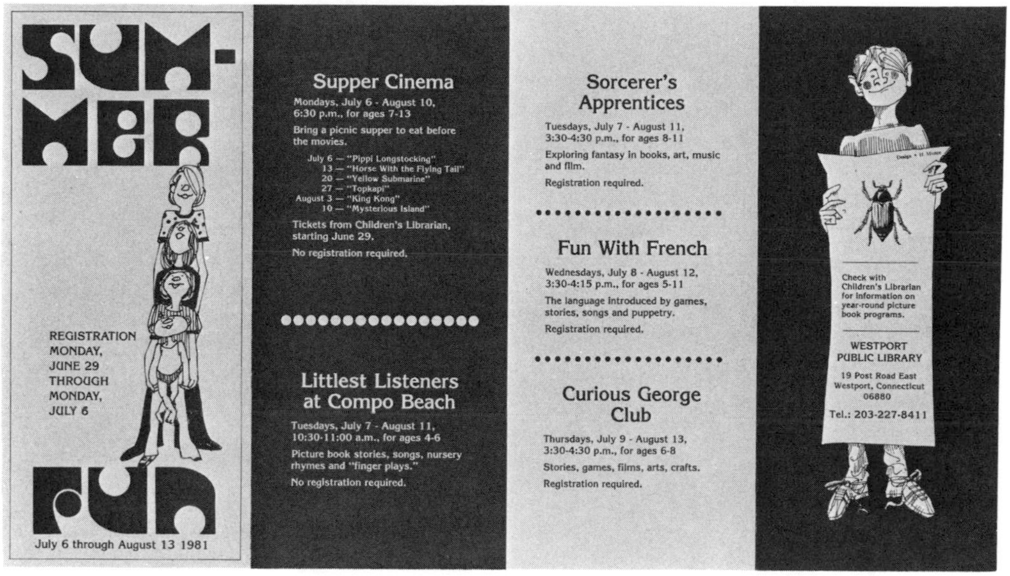

Often necessity forces you into creating different forms. This one came about because the budget allowed only for printing on one side—and at the same time it had to fold to fit a pocket and still have a front cover. The reverse side, of course, is blank.

Folds

Among the many other decisions you have to make in the design of a mailing piece, a brochure or an announcement, is the form it's to take.

Often your decision will be dictated by the amount of material the piece must contain.

At other times, with scant material and a generous budget, you can be more expansive and make greater use of that luxury all designers crave: white space.

Here are a few examples of folds and forms that are available to you at all job-printers and at many quick copy centers.

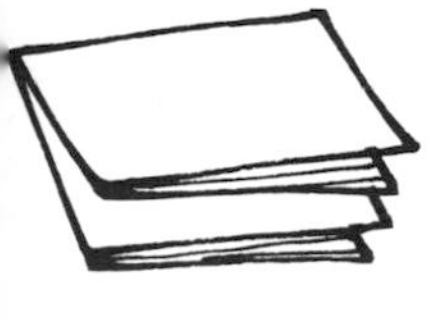

French fold

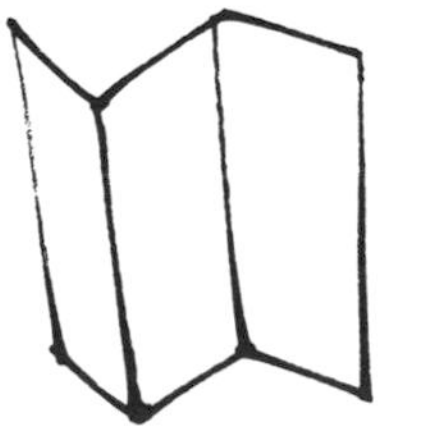

Accordion fold

Pocket size

Self-mailer

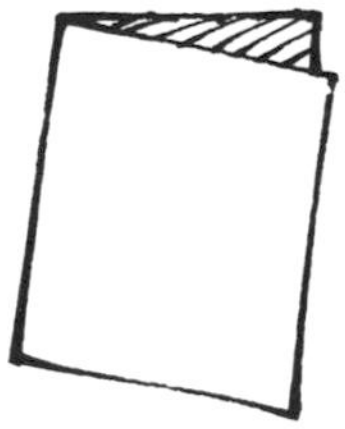

Simple single fold

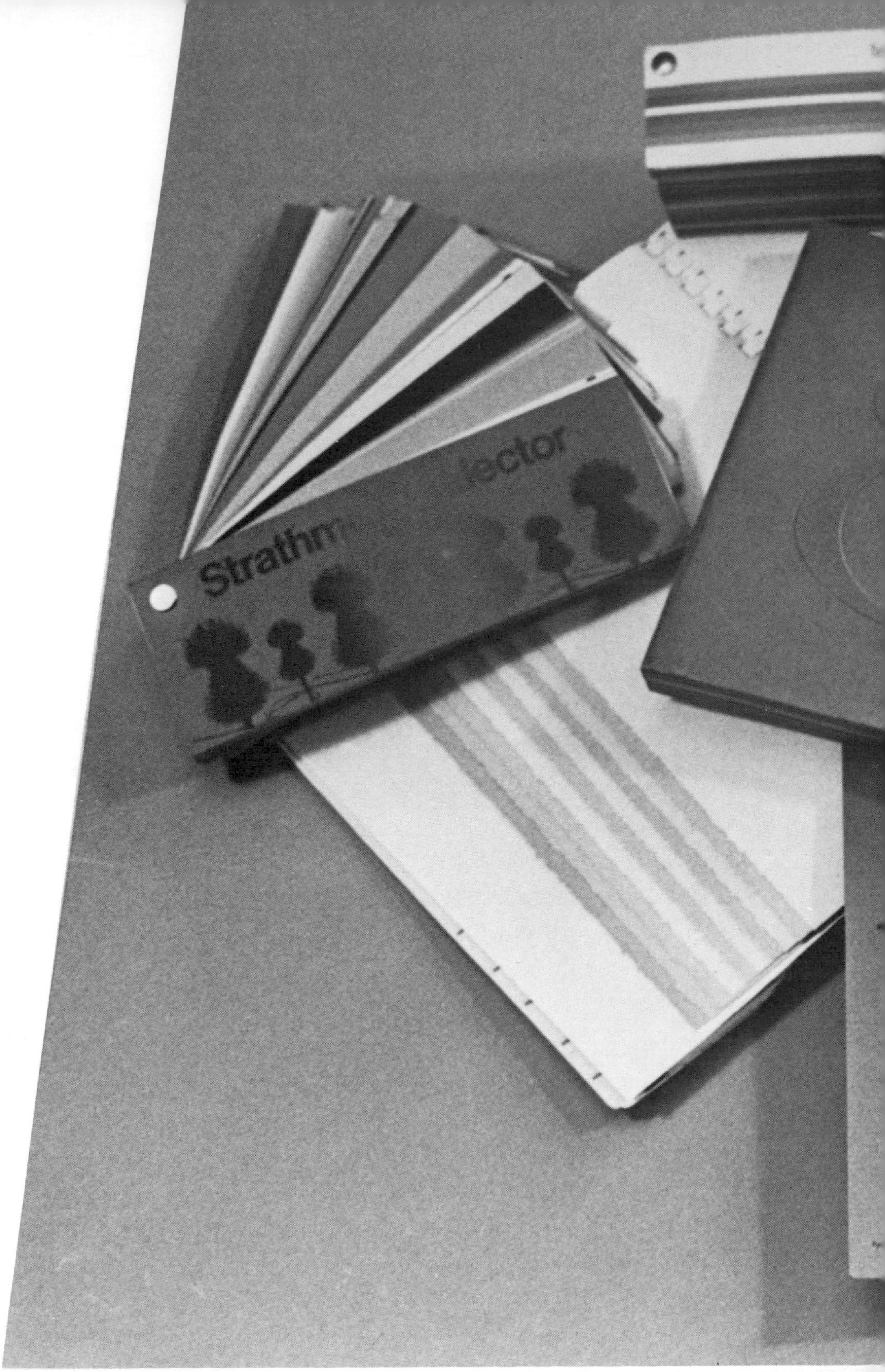

Here are several sample books from different paper companies. They allow you to see the color, heft the weight and feel the texture.

Photo by John D. Brewster

Picking the paper for a job is a pleasurable one—it has all the qualities of shopping for gifts or studying a menu: many desirable choices followed by a tough decision.

Standard Old Colony envelope styles and sizes

Over the years, the graphic arts industry and envelope converters have standardized on certain envelope styles. And within each style, a range of standard sizes has evolved.

Each style tends to carry a general type of printed matter, i.e., commercial and official envelopes are designed for letterheads, invoices, statements and other general business correspondence. Further, the different sizes offered in that style are each designed for a specific need within that general area of usage — i.e., No. 10 Official for 8½ x 11 letterheads.

It has been estimated that standard styles and sizes of envelopes carry over eighty percent of the nation's mail. Little wonder there's hardly a standard style/size envelope that Old Colony doesn't either stock or can't supply very quickly. Because they're so readily available, standard envelopes eliminate the more expensive and time consuming process of waiting for a made-to-order envelope. Consequently, **they should always be one of the starting points in planning any mailing.**

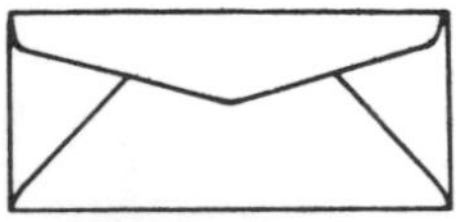

Commercials/Officials — for a host of business and personal correspondence purposes, i.e., letterheads, invoices, statements, writing stationery, direct marketing mailings, etc.

6¼ Commercial	3½ x 6
* 6¾ Commercial	3⅝ x 6½
8⅝ Official	3⅝ x 8⅝
7 Official	3¾ x 6¾
7¾ Official	3⅞ x 7½
* Monarch	3⅞ x 7½
* 9 Official	3⅞ x 8⅞
* 10 Official	4⅛ x 9½
11 Official	4½ x 10⅜
12 Official	4¾ x 11
14 Official	5 x 11½

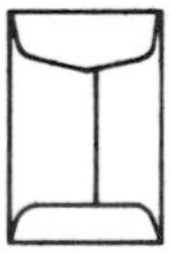

Coins — a multi-purpose envelope for small parts, coins and currency, etc.

1	2¼ x 3½
2 Pay (OS)	2½ x 4¼
* 3	2½ x 4¼
4	3 x 4½
4½	3 x 4⅞
5	2⅞ x 5¼
5½	3⅛ x 5½
6	3⅜ x 6
7	3½ x 6½

* Most popular sizes.

Booklets — for annual reports, brochures, sales literature and manuals. Used in volume mailings processed by inserting and sealing equipment.

3	4¾ x 6½	7½	7½ x 10½
5	5½ x 8⅛	* 9	8¾ x 11½
6	5¾ x 8⅞	* 9½	9 x 12
6½	6 x 9	10	9½ x 12⅝
6⅝	6 x 9½		

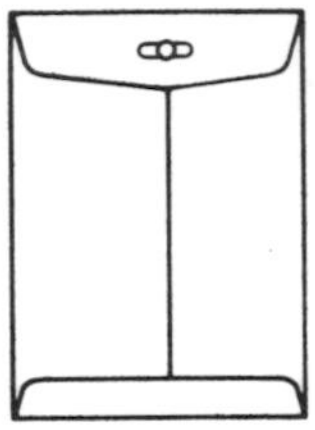

Clasps — the basic catalog envelope with a metal clasp.

5	3⅛ x 5½	80	8 x 11
10	3⅜ x 6	83	8½ x 11½
15	4 x 6⅜	87	8¾ x 11¼
11	4½ x 10⅜	* 90	9 x 12
25	4⅝ x 6¾	93	9½ x 12½
35	5 x 7½	94	9¼ x 14½
14	5 x 11½	95	10 x 12
50	5½ x 8¼	* 97	10 x 13
55	6 x 9	98	10 x 15
63	6½ x 9½	105	11½ x 14½
68	7 x 10	* 110	12 x 15½
* 75	7½ x 10½		

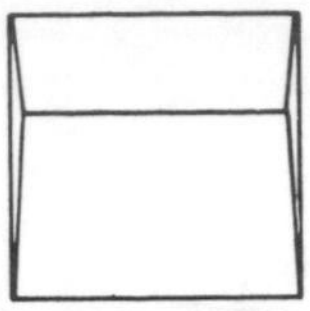

Squares — with square flaps, for announcements, booklets, promotions, etc.

* 6½	6½ x 6½	* 8½	8½ x 8½
7	7 x 7	9	9 x 9
7½	7½ x 7½	9½	9½ x 9½
8	8 x 8	10	10 x 10

* Most popular sizes.

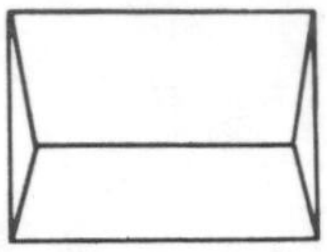

Square Flap A-Sizes — for announcements, small booklets, brochures or promotional pieces, and more recently for distinctive business stationery.

* A-2	4⅜ x 5¾	A-8	5½ x 8⅛
* A-6	4¾ x 6½	* A-10	6 x 9½
A-7	5¼ x 7¼	A-Long	3⅞ x 8⅞

Fitting standard Old Colony envelopes to their enclosures (and vice versa)

In the case of a frequently used enclosure, the style and size of the envelope is virtually a foregone conclusion. As noted in the previous section, an 8½ x 11 letterhead almost always is mailed in a No. 10 Official envelope (even though No. 9's and No. 10 Square Flaps are sometimes used).

Frequently, however, the enclosure doesn't predetermine an envelope choice so readily. In that case, you have two choices.

First, you can go ahead and concept the enclosure, even produce it in its entirety, and then look for the envelope. If your experience is typical, however, your enclosure may well wind up requiring a special-making envelope, a purchase process that costs you extra time and money.

Or, you can save that time and money by going to Old Colony's listing of standard styles and sizes right at the very outset and tailor-make your enclosure to fit a standard envelope style and size.
Important: When following this latter procedure, keep in mind — there are a number of factors that affect enclosure size, factors that should be considered right at the very outset. Such things as enclosure bulk, method of inserting, sealing, and postage metering, for instance.

For inserting by hand, specify an envelope opening ⅛" to ¼" wider than the enclosure, and the overall envelope length ¼" to ⅜" longer than the enclosure. As part of your design process, be sure to check out your specs by building a dummy enclosure with an envelope, using paper and a matching envelope easily obtained from your paper merchant.

For machine inserting, enclosures and envelopes should both be compatible with the mechanical systems used by your printer or mailing house. Ask them to provide specifications, which should then be followed exactly.

Here are recommended enclosure dimensions for today's most popular envelope sizes.

Commercials/Officials

Item	Envelope Size	Single card or folded form Enclosure Size
6¾	3⅝ x 6½	3½ x 6¼
Mon	3⅞ x 7½	3¾ x 7¼
9	3⅞ x 8⅞	3¾ x 8⅝
10	4⅛ x 9½	4 x 9¼

It may seem silly to remind you to be sure to double (and triple) check the final size of a mailing piece to be certain it will fit into a stock envelope. Take it from one who has sinned, it's very important!

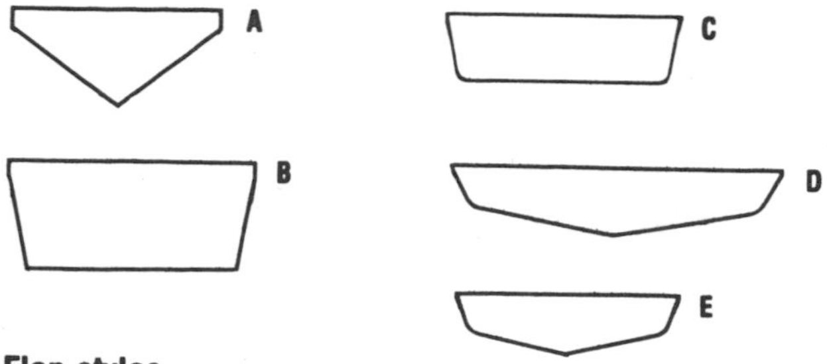

Flap styles

Envelopes are further designated by the style of the seal flap.

A — Pointed
B — Square
C — Wallet
D — Press-Ready®
E — Mailpoint

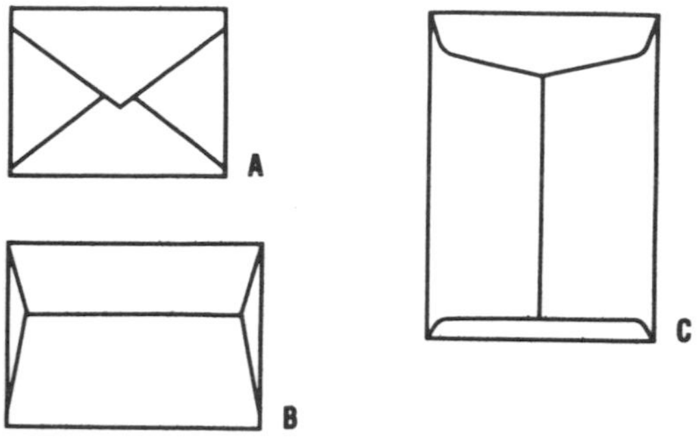

Seam styles

Within each general designation, styles are also determined by location of the seams.

A — Diagonal Seams: start at the corners and run diagonally to the center. Used commonly in correspondence commercials and pointed flap announcements.

B — Side Seams: seams at sides only, leaving large uninterrupted space on front and back for printing. Booklets and square flap announcements utilize this design.

C — Center Seam: located in center of envelope. This design is used when an extremely rugged envelope is needed. Consequently, it is often made from heavier papers i.e., strong krafts.

Adhesives and closures

Different grades of paper require different gums. Old Colony's gum formulations, including dextrine, resin and other adhesive materials, are the best in the industry. Gum qualities are constantly evaluated for adhesion, cohesion, acidity and other factors to prevent warping, discoloration, separation or tacking down of top flaps.

Old Colony's various adhesives and methods of closure are:

BACK GUM — for permanently sealing seams.
SEAL GUM — applied to seal flaps for moistening when flaps are to be sealed.
SIMPLE-SEAL® — self-adhering latex adhesive on two surfaces that seal on contact.
CLASP — double pronged metal clasp for additional security.
BUTTON AND STRING — string tie-down for repeated usage.

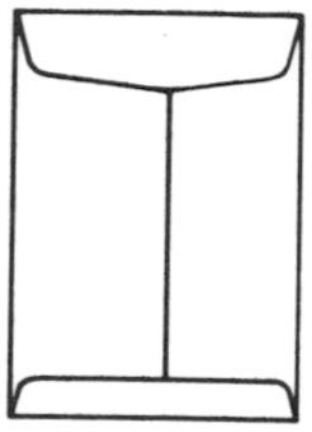

Catalogs — for catalogs, large booklets and other heavy enclosures usually inserted by hand.

7 Glove	4 x 6⅜
8 Glove	3⅞ x 7½
10 Policy	4⅛ x 9½
11 Policy	4½ x 10⅜
14 Policy	5 x 11½
1 Scarf	4⅝ x 6¾
3 Scarf	5 x 7½
4¼ Scarf	5½ x 7½
6 Scarf	5½ x 8¼
* 1 Catalog	6 x 9
* 1¾ Catalog	6½ x 9½
3 Catalog	7 x 10
6 Catalog	7½ x 10½
8 Catalog	8¼ x 11¼
* 9¾ Catalog	8¾ x 11¼
* 10½ Catalog	9 x 12
12½ Catalog	9½ x 12½
* 13½ Catalog	10 x 13
14½ Catalog	11½ x 14½
15½ Catalog	12 x 15½

Air or First Class Mailers — booklet style envelopes with red and blue or green diamond borders to expedite postal handling.

1st Class	9 x 12
1st Class	9½ x 12½
Air	9½ x 12½
1st Class	10 x 13

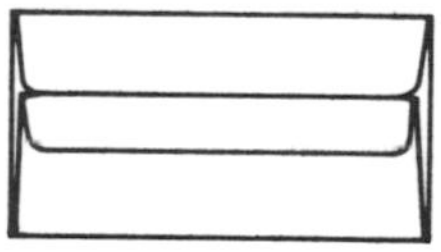

Simple-Seal® — with two latex surfaces that seal on contact. Efficient and convenient for small businesses, professional people, personal stationery.

6¾ Commercial	3⅝ x 6½
Monarch	3⅞ x 7½
9 Official	3⅞ x 8⅞
10 Official	4⅛ x 9½

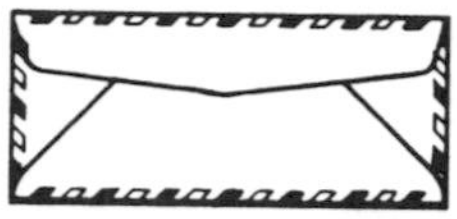

Air Mail — commercial/official envelopes made from lightweight papers, with red and blue borders and air mail legends.

6¾ Official	3⅝ x 6½
Monarch	3⅞ x 7½
9 Official	3⅞ x 8⅞
10 Official	4⅛ x 9½

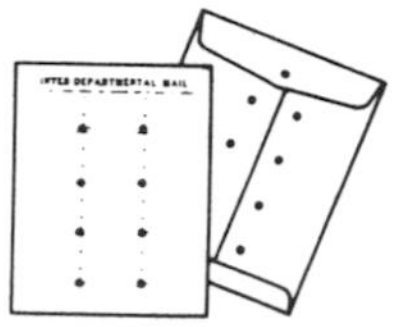

Interdepartmental — to speed inter-office movement of correspondence and material, with eight holes and printed route listing, plain or with button & string.

10 x 13

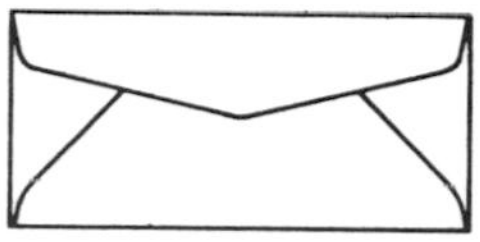

Punch Card Accounting — for computer style forms; punch cards, checks, invoices, statements, etc.

953	3½ x 7⅝
940	3⅝ x 7½
954	3⅝ x 7⅞
814	3⅝ x 8⅜

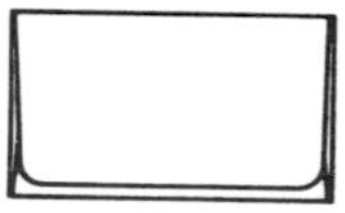

Combination Statement & Return — Dual-purpose envelope that includes a statement and provides for the return of orders or remittances.

6¼	3½ x 6
6½	3½ x 6¼
6¾	3⅝ x 6½

Thank You! — an expression of appreciation, or any special message, is seen through a SEE-CLEAR® window, reverse printed

6¾	3⅝ x 6½
10	4⅛ x 9½

Stock envelopes, like people, come in many sizes and shapes: these two pages from the "Old Colony Specifier," published by the Old Colony Envelope Co. of Westfield, MA, will give you an inkling of what's available.

REPRODUCTION:
Wherein you deal with your local printer & learn the facts of his life & machines whilst seeing your handiwork come to fruition.

When you've finished your mechanical with everything accurately in place—all instructions clearly marked, all finished art in the package—your job is "Camera Ready," and you're off to the printer. Everything you can control is now done, except for seeing a blueprint of it trimmed to size and folded. This is your last whack at it before it gets printed. When you're satisfied, you initial it, say a short prayer and await delivery.

If it's a complicated and expensive job, you may awaken trembling in the night and pray that you haven't goofed. Relax, if you can. Know that everywhere there are other designers sweating out their jobs, too. Moist palms go with the territory.

Depending on where you live, you'll either have a choice of a couple of printers or no choice. I hope there's more than one available to you so you can shop for the best price on small jobs, and also shop for the one where you feel most comfortable and where you get the most help and attention.

Make sure you ask questions of the printer you deal with. Make it a point to come away better informed after each visit.

Also, look around at the printed samples of jobs he displays. See how other people did it. Note color combinations, the use of stocks, the sizes of type, the folds and forms. Note everything and ask, ask, ask.

It's possible to reproduce graphic images from something as simple as a rubber stamp to a million dollar high-speed press. Between these two extremes and many processes is the commonest one, the one you'll most likely be dealing with; it's called *Offset.* In its simplest terms it works by literally "setting off" an image from the press's inked rubber rollers (transferred from the zinc plate) onto the paper.

When you get to know your printer, ask him to show you the several simple steps that take place from the time you deliver your material to him until it gets on press, etched on the zinc sheet. It's ingenious and fascinating—and simple—and all based on the age-old principle that oil and water don't mix.

STRIP-IN GALLERY LOGO

Thirty-three exhibitions and two-thousand pieces of art later, The GE Gallery celebrates its fifth anniversary.

Since its first exhibition, which coincided with the opening of the new Corporate Headquarters in 1974, the Gallery has displayed an endless variety of art in all forms. Its shows have ranged from traditional oil and watercolor depictions of nature to examples of the avant-garde in American fine arts. From pure craft, such as the historical quilt exhibit that celebrated the Bicentennial, to the dazzling display of treasures from the New Britain Museum of American Art that helped us observe GE's 100th Anniversary, the Gallery's exhibits have brought viewers the broadest spectrum of today's art and in every conceivable material from advertising assemblages to sculptured works in all mediums. The art experiences it has opened up have ranged from super realism to total abstraction, from surrealism and personal fantasy to pictorial fun.

For those of you who never miss the exhibitions, you'll be pleased to know that another full schedule of six new shows has been planned for the next year. And for those of you who are new to GE or have not availed yourself of the opportunities the Gallery provides, do take a look and discover what you've been missing.

Stevan Dohanos Gallery Director

Howard Munce Gallery Coordinator

GENERAL ELECTRIC

300 COPIES \ 80 lb. COVER (wh.)

BLACK INK \ (TORINO)

60% Tint

Dolli

Tingle

Ward

Brackett

A joint exhibition of paintings

60%

August-October

These instructions are on acetate for demonstration—normally they'd be done on a tissue overlay.

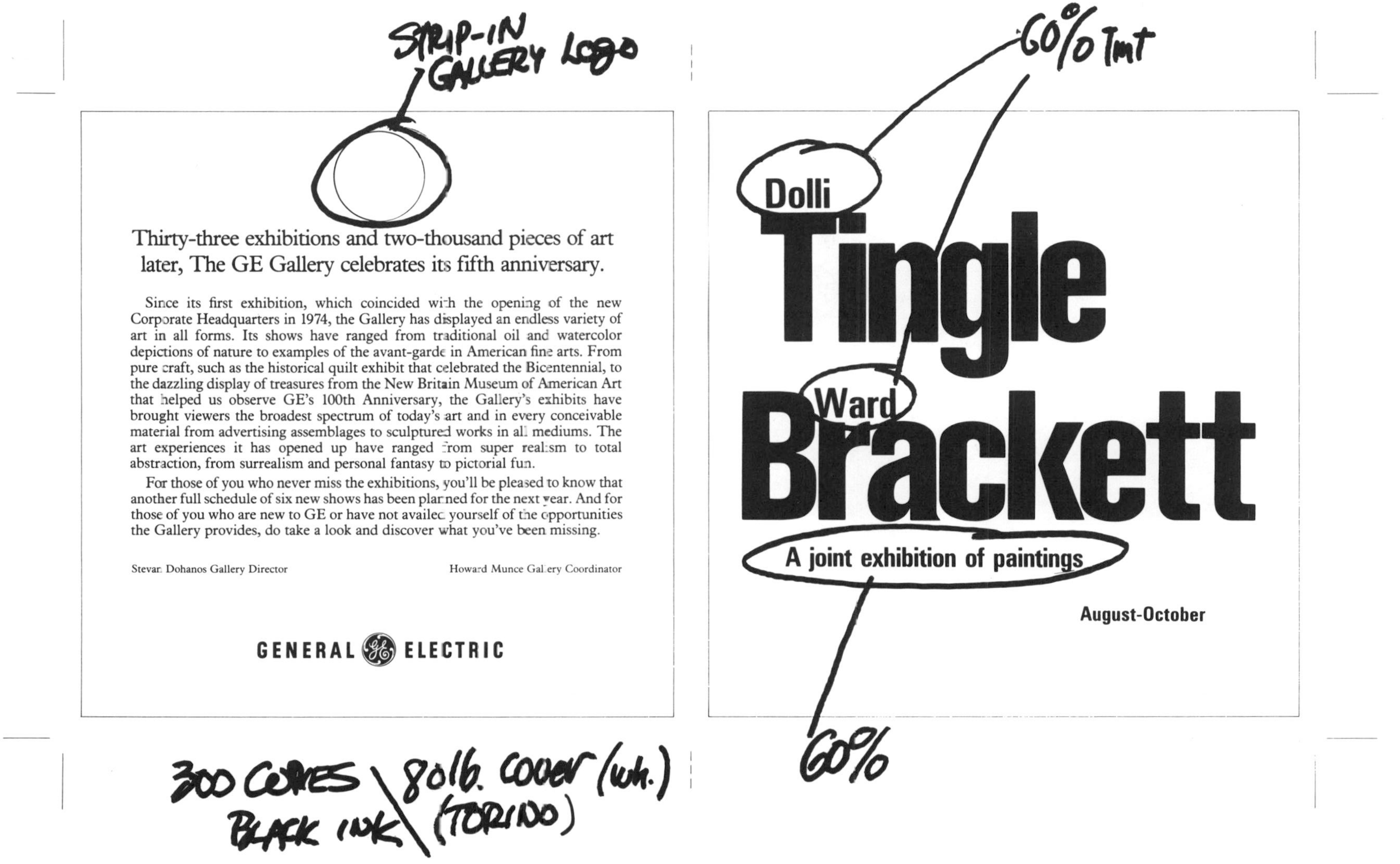

THE NEIGHBORHOOD COPY CENTER

The neighborhood copy center is now a part of our everyday life. They are a phenomena that has sprung into being because of the amazing advances in duplicating technology. I'm referring to the so-called "xerox" which has become a generic term much to the consternation of the Xerox Corporation, since so many others also manufacture similar equipment.

Copy centers now have advanced equipment that allows for the use of other than chemically-treated paper. So it's now possible to bring your own paper for them to print your job on. This is a nice new turn, since the center's paper supply is usually limited to common colors and ordinary textures.

Make sure that your camera-ready art is perfectly clean and that you avoid cementing thick pieces of paper to the mechanical. The new machines are so sensitive that they'll pick up all foreign matter such as rubber cement. And thick additions cast a shadow that produces an unwanted gray mark.

Copy centers have a definite price list for every step—so you'll know just where you stand at all times.

Many copy centers now have offset printing equipment in addition to their duplicating machines. This means you can produce much more sophisticated work. Check out the capabilities, prices and service between your center and your local job-printer.

In some cases they will both have typesetting and printing capabilities. If they don't, then you must go to a typographer for your type. Then put it in mechanical form—camera-ready. The big advantage of a place that does both type *and* printing, is that you can make last minute type changes right in the same place. There are hairy times and deadlines when that can be your salvation.

Though copy centers have limitations, they are a welcomed addition in the graphic field. They're excellent for short runs, simple production, fast service and great price.

These are some typical jobs done at a full-fledged printing establishment and at a local copy center.

Baker Graphics

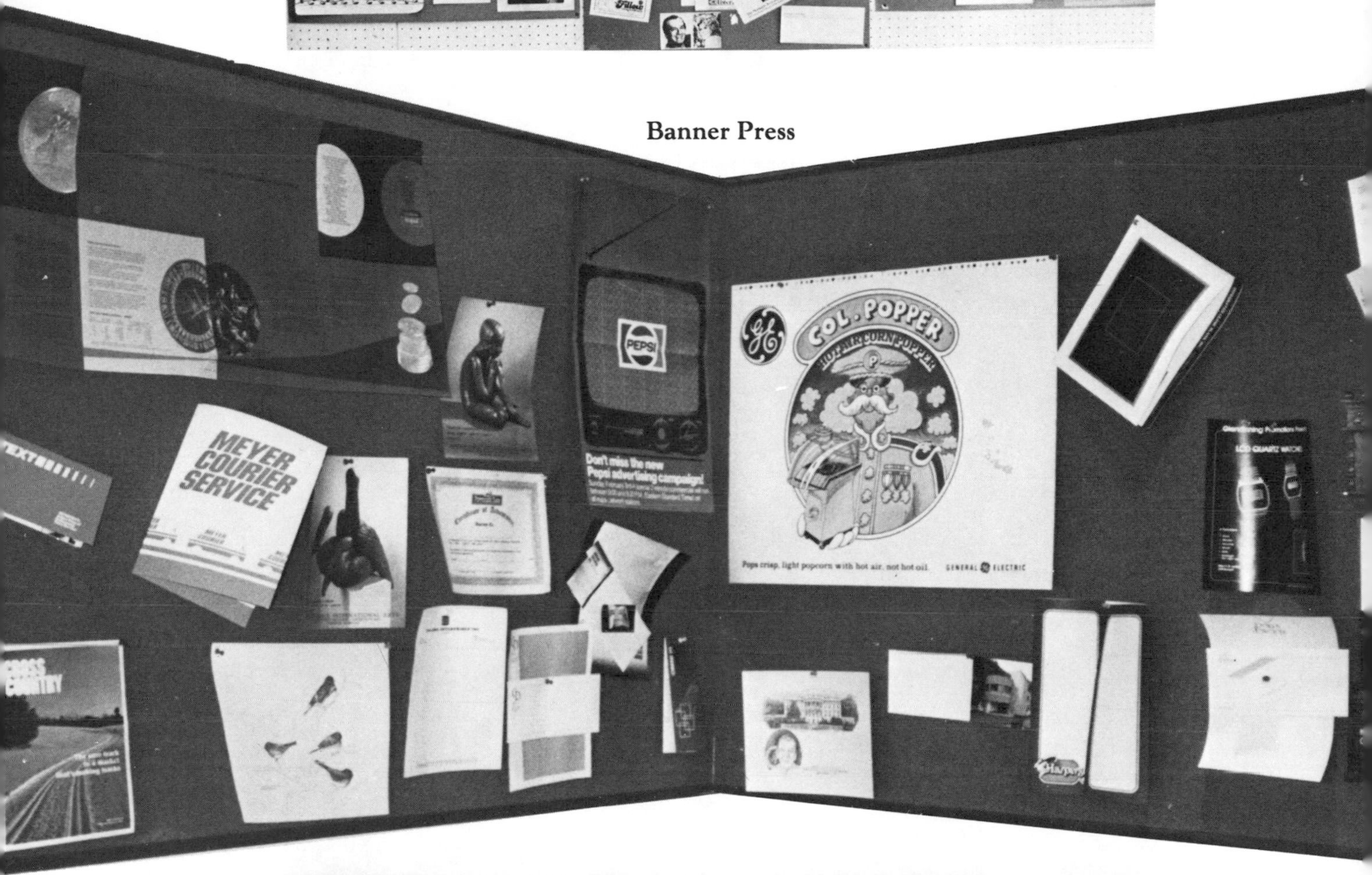

Banner Press

Klein's Printing

DESIGN:

Showing several modest examples of the work of the author whose aim in presenting & explaining them is to try to impress upon you the value of idea & simplicity.

It would be a hopeless task to try to teach the subject of design in so small and modest a book as this. Only a few simple principles can be covered here. But they're ones that are readily grasped.

I felt it more effective if, instead of talking design in theory, I could better make my points by example. And to help you see the merits of the examples I've chosen, I've added comments and offered guidance.

In all cases I've picked examples that have Idea, Simplicity and Taste. None of them are Tricky, Ornate or Overdone. If you'll keep those six words in mind and strive for the first three and resist the latter three you'll produce creditable work.

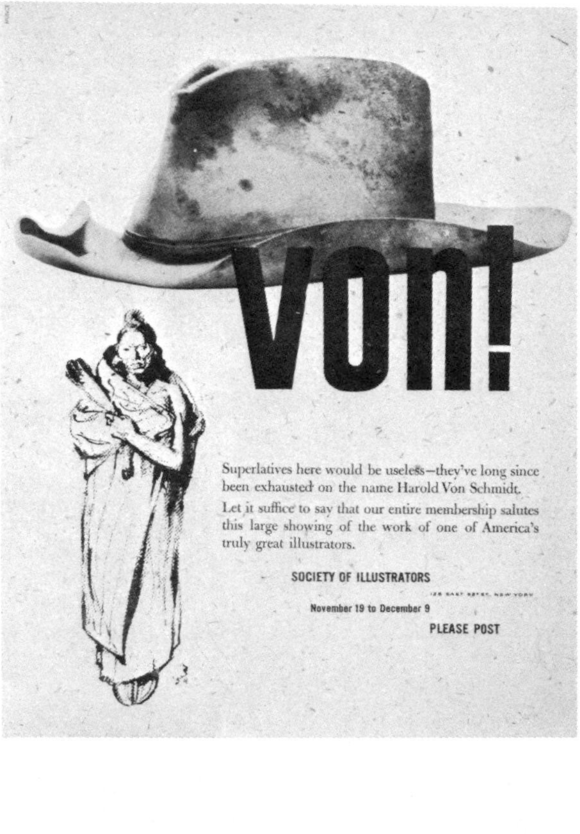

Von!
Superlatives here would be useless—they've long since been exhausted on the name Harold Von Schmidt.
Let it suffice to say that our entire membership salutes this large showing of the work of one of America's truly great illustrators.
SOCIETY OF ILLUSTRATORS
November 19 to December 9
PLEASE POST

RUDY ZALLINGER

July-September/1981

Rudolph Franz Zallinger was born November 12, 1919 in Irkutsk, Siberia. His father, Franz, was an Austrian, an artist drafted into the army in World War I, who after a year of trench fighting was captured on the Russian Front and sent to prison camp in Siberia. In due course his entrepreneurship and resourcefulness gained him enough freedom to meet and marry Marie Koncheravich, the daughter of a Polish civil engineer who was working on the Trans-Siberian Railway.

When their child Rudolph was nine-months old, the Zallingers began a long-drawn-out emigration via Harbin, Manchuria to the United States. They settled in Seattle. As a child, he began learning about art from his father and in school and from private teachers. He reflects, "I do not remember just when I first started drawing and painting. I think I learned how to wash my father's paint brushes before I could brush my teeth. Furthermore, I recall that I was ten years old when I entered a life class." And, by then he had sold his first painting, for five dollars — a small child's fortune!

Also a serious student of the piano, at seventeen having finished high school, he was faced with a hard decision as to primary focus. After agonizing over it, the choice was made to enter art school, at that point the University of Washington. However, as usual the fates intervened as a visiting artist from whom Rudolph was awarded a study scholarship for that summer, persuaded him to seek enrollment at Yale. He hurriedly applied and was accepted for that Autumn of 1937. Thereupon, he entered the Yale School of the Fine Arts in the five-year Bachelor of Fine Arts program in painting. He won a merit scholarship in every term, after the first.

At the beginning of his senior year, after marrying a classmate, Jean Day (now an illustrator of over 70 books), Zallinger was awarded the commission to execute a mural in the Great Hall of Reptiles in the Peabody Museum of Natural History at Yale. This work was to be the all-consuming effort on the part of the artist, in collaboration with the fine curatorial staff of scientific specialists. The project of research, design and execution as a secco-fresco, 110 feet in length and 16 feet in height spanned a five-year period. He was teaching at his alma mater during all that time, two days a week, but otherwise his energies were mainly channelled in this project. It was completed in June, 1947. In May of 1949, the Pulitzer Committee granted the artist the Pulitzer Award in Painting for 1949, mainly but not exclusively, for the Peabody work.

He resigned from Yale in 1950 and moved the family to Seattle and launched upon a difficult career of free-lance work of various sorts, decorative panels, small murals, portraits and easel-paintings and the like. Late in 1951, an editor from Life Magazine telephoned to describe to him the proposed "World We Live In" series. It was to be the most comprehensive ever attempted in the history of general publication. Life wanted to use the Dinosaur mural, and proposed to have him produce a similar treatment for the evolution of mammals. Zallinger returned to Peabody, and following the same procedure produced a comparable panel which was reproduced in the Series. Eventually, a second large mural, a secco-fresco 60 feet by 6 feet was produced in the same manner as the Dinosaurs, completed in November 1967.

In between, he illustrated books, magazine articles, painted portraits and in 1971, commenced a series of landscape paintings — being essentially a naturalist by primary inclination. All that time he taught drawing and painting at the University of Hartford and some time at the Paier School of Art, Hamden, Ct. A retrospective exhibition consisting of about 100 pieces was mounted at the New Britain Museum of American Art in February-March 1979.

In January of 1980, he was awarded an honorary Doctor of Fine Arts by the University of New Haven. The citation reads in part: "Your great murals at the Peabody Museum have been widely acclaimed for their artistic and scientific integrity. From your palette has come a rich varity of subjects dealing with time past and present. Common to all your work is a perceptive realism founded on painstaking research and rendered with the acuity of the creative and disciplined artist. Your paintings widen the horizons of those who see them, offering experiences ranging from the ephemeral, limpid surfaces of running water to the firm contours of mountain slopes, from the life forms of the rain forests to the leaders of academia. For your accomplishments as artist and teacher of artists, the University is honored . . . ."

On February 29, 1980, Zallinger became the first non-scientist to receive the Addison Emery Verrill Medal which is awarded by the curators and trustees of the Yale Peabody Museum, "to honor some signal practioner in the arts of natural history and natural science. Yale president A. Bartlett Giamatti presented the medal calling the natural history murals "a fusion of scientific accuracy and artistic genius. Guided by your own diligent research and painstaking collaboration with scientists, your imagination has allowed us a glimpse into the past worlds that no human eye ever witnessed. It was your innovation to blend the static frames of successive geologic ages into grand panoramas that swept through time, capturing the dynamic force of life as it evolved . . . ."

Of the landscape paintings which numerically comprise the larger part of his work, Roger Tory Peterson, the renowned ornathologist and artist writes, "The Zallinger paintings are a timeless record radiating a distinct sense of nature's creative forces."

Stevan Dohanos, Gallery Director

Howard Munce, Gallery Coordinator

The copy for this announcement was very long, but fascinating and good. So the decision was made not to cut it. Hence, I wanted to have a book-like quality. And since the copy occupied so much space I decided to let the artist's intriguing name be the only illustration.

This strong round letter (Bauhaus) had a right feel and I was tickled to discover the fortunate break in the similarity of the "U" and the "A" above and below each other. The dates were placed in the upper right corner to visually complete a rectangle.

One of the reasons for picking this text type (Garamond) was that it comes in a bold version, too. So I was able to start the text with his full name bold and black, then go to the regular weight.

The inside spread is all type. It continues the information begun on the front cover.

The GE Gallery presents collages and assemblages by Fred Otnes

Illustrator Fred Otnes builds his pictures image by image, with perfect control and with gratifying results.

Before beginning work on a particular collage, he immerses himself in the subject. He researches and reads all the available background material, selects pertinent cut-outs from among "hundreds of thousands of pieces of paper"—pictures, photos, and illustrations from periodicals that he has clipped and saved for years. Then the cut-outs are arranged and rearranged until, by means of superimposition and juxtaposition, a complex density of detail is achieved that produces a unified collage.

Otnes' careful mastery of his work has won him over 100 awards for clients such as GE, Exxon, RCA, General Motors, ABC-TV and American Express, and his illustrations have appeared in Fortune, Life, Look, Journal, Lamp, Sports Illustrated, Seventeen, Saturday Review, Lithopinion and McCall's.

Born in Junction City, Kansas, Otnes attended the Chicago Art Institute and the American Academy. He presently resides in West Redding, Connecticut.

This strong, punchy photo existed, so it was up to me to match the other elements, type and format, to it to enhance its contemporary look and to stay in sympathy with the actual exhibition it touted and the eminent artist represented.

"Otnes" is a distinctive name, and spelled out in this phototype face (Cinema Shaded) a happy happenstance caught my eye—the full circle of the "O" and the "E" began and visually concluded the word (the final "S" is so slim it doesn't matter).

The inside fold employs a simple device that gives you a lot for your money. When there's a great deal of positive paper showing, and a section with rather simple elements on it, I like to do as I've done here, reverse the elements so that the paper comes through as the "color" against the negative background. Note how different, how luminous the name Otnes looks in reverse.

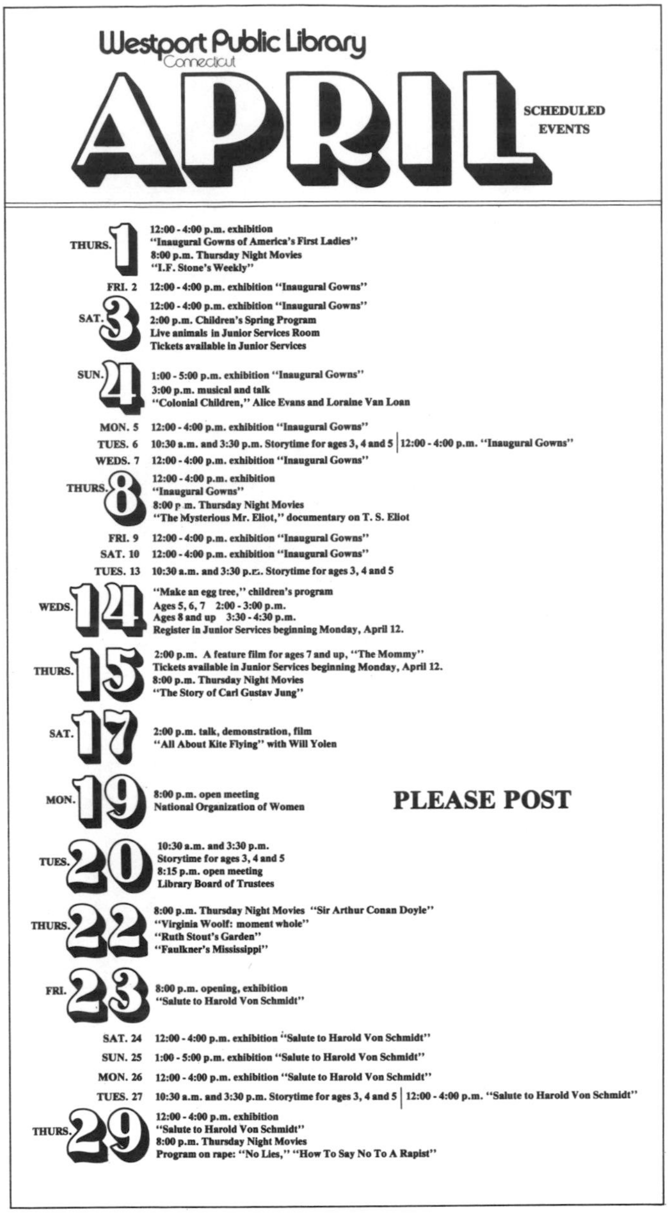

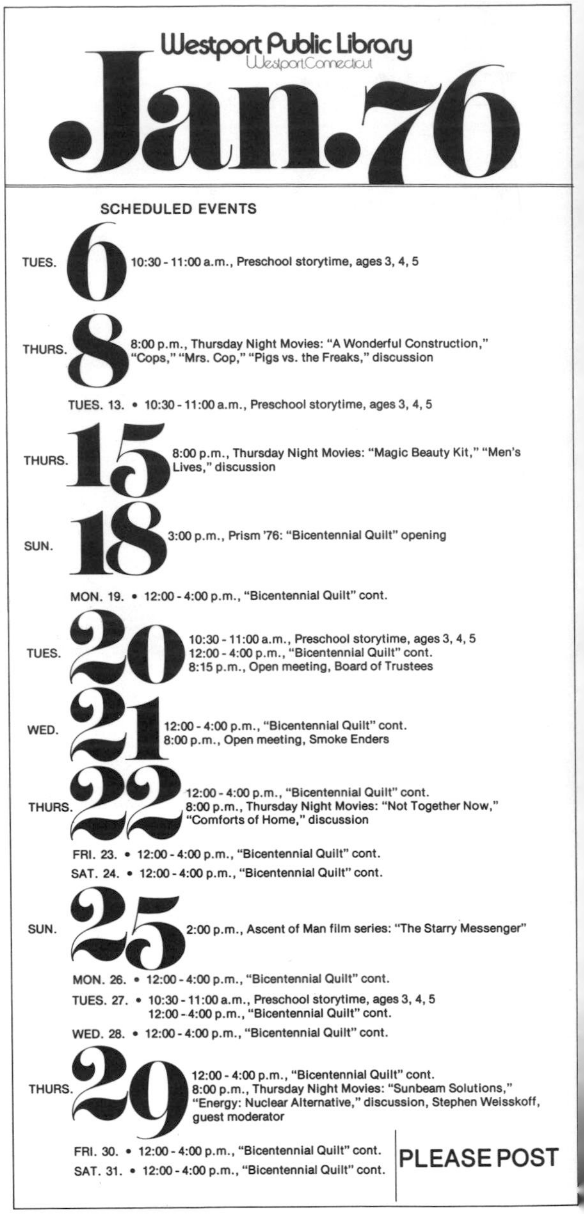

Here are four inexpensive posters designed to take advantage of as many money-saving devices as possible.

The large display type and numerals are enlarged from old type proofs from my files.

The size and shape of each sheet is the maximum that will fit the small offset press. Larger presses add considerably to the cost.

Westport Public Library
SCHEDULED EVENTS
June, July,
August '76

June

MON. 7 8:00 p.m. Open Meeting, Friends of Westport Library
SAT. 12 7:00 - 10:00 p.m. Opening, exhibition, "Westport Artists of the Past"
MON. 14 - 11·00 a.m. - 8:00 p.m., exhibition, "Westport Artists of the Past"
TUES. 15 - 11:00 a.m. - 5:00 p.m., exhibition, "Westport Artists of the Past"
8:15 p.m. Open meeting, Library Board Trustees
WEDS. 16 - 11:00 a.m. - 5:00 p.m., exhibition, "Westport Artists of the Past"
THURS. 17 - 11:00 a.m. - 8:00 p.m., exhibition, "Westport Artists of the Past"
FRI. 18 11:00 a.m. - 5:00 p.m., exhibition, "Westport Artists of the Past"
8:30 p.m. Talk: Edgar Mayhew, Director, Lyman-Allyn Art Museum
SAT. 19 - 10:00 a.m. - 4:00 p.m., exhibition, "Westport Artists of the Past"
MON. 21 - 11:00 a.m. - 8:00 p.m., exhibition, "Westport Artists of the Past"
TUES. 22 - 11:00 a.m. - 5:00 p.m., exhibition, "Westport Artists of the Past"
WEDS. 23 - 11:00 a.m. - 5:00 p.m., exhibition, "Westport Artists of the Past"
THURS. 24 - 11:00 a.m. - 8:00 p.m., exhibition, "Westport Artists of the Past"
FRI. 25 - 11:00 a.m. - 5:00 p.m., exhibition, "Westport Artists of the Past"
SAT. 26 - 10:00 a.m. - 4:00 p.m., exhibition, "Westport Artists of the Past"
MON. 28 - 11:00 a.m. - 8:00 p.m., exhibition, "Westport Artists of the Past"
TUES. 29 - 11:00 a.m. - 5:00 p.m., exhibition, "Westport Artists of the Past"
WEDS. 30 - 11:00 a.m. - 5:00 p.m., exhibition, "Westport Artists of the Past.

July

SAT. 3 2:00 p.m., film, "A Connecticut Yankee in King Arthur's Court"
SAT. 10 3:00 p.m., contest, handcrafted clothing "American Denim"
MON. 12 10:00 a.m., Demonstration of hand and punch rug hooking with Marjorie Wall
TUES. 13 - 4:00 p.m. - 4:30 p.m., Storytime for Juniors
WEDS. 14 - 3:30 p.m. - 4:30 p.m., Reading Club for Juniors, grade 2 and up. Registration required.
THURS. 15 - 4:00 p.m. - 4:30 p.m., Movies for Juniors
TUES. 20 - 4:00 p.m. - 4:30 p.m., Storytime for Juniors
WEDS. 21 - 3:30 p.m. - 4:30 p.m., Reading Club for Juniors, grade 2 and up. Registration required.
THURS. 22 - 4:00 p.m. - 4:30 p.m., Movies for Juniors
TUES. 27 - 4:00 - 4:30 p.m., Storytime for Juniors
WEDS. 28 - 3:30 - 4:30 p.m., Reading Club for Juniors, grade 2 and up. Registration required.
THURS. 29 - 4:00 p.m. - 4:30 p.m., Movies for Juniors

August

TUES. 3 - 4:00 - 4:30 p.m., Storytime for Juniors
WEDS. 4 - 3:30 p.m. - 4:30 p.m., Reading Club for Juniors, grade 2 and up. Registration required.
THURS. 5 - 4:00 p.m. - 4:30 p.m., Movies for Juniors
TUES. 10 - 4:00 p.m. - 4:30 p.m., Storytime for Juniors
WEDS. 11 - 3:30 p.m. - 4:30 p.m., Reading Club for Juniors, grade 2 and up. Registration required.
THURS. 12 - 4:00 p.m. - 4:30 p.m., Movies for Juniors
TUES. 17 - 4:00 p.m. - 4:30 p.m. - Storytime for Juniors
WEDS. 18 - 3:30 - 4:30 p.m., Reading Club for Juniors, grade 2 and up. Registration required.
TIIURS. 19 4:00 p.m. - 4:30 p.m., Movies for Juniors

The Library will be closed on Sunday
during June, July and August.

PLEASE POST

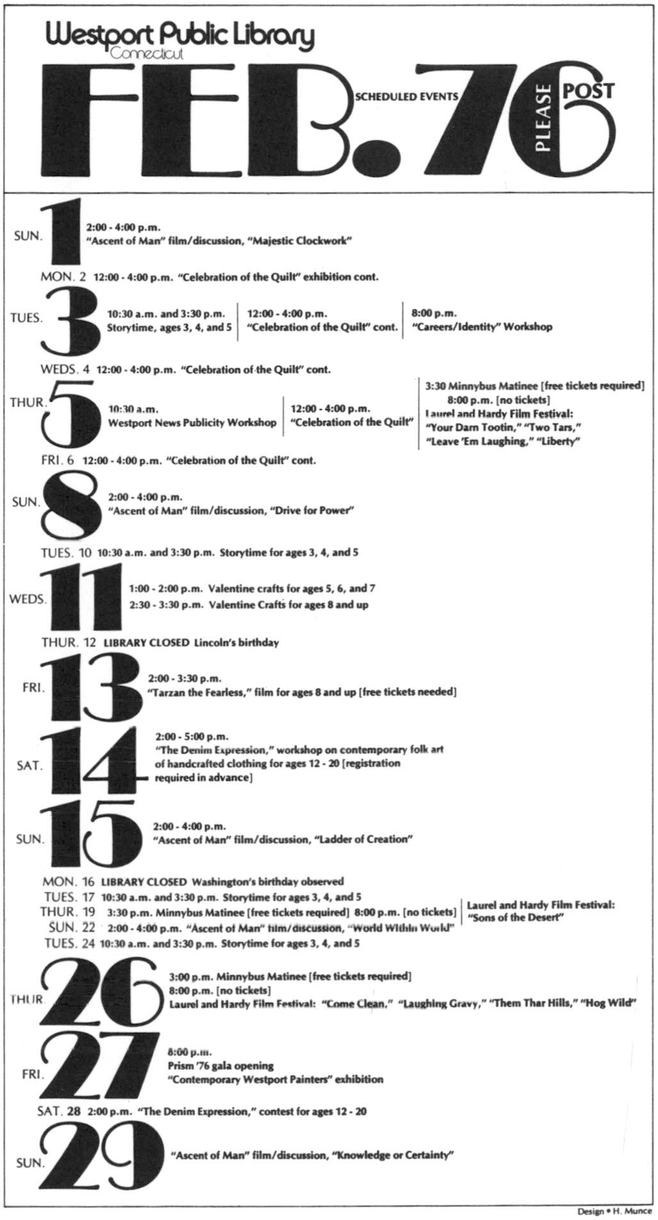

Westport Public Library
Connecticut
FEB. 76 SCHEDULED EVENTS PLEASE POST

SUN. 1 2:00 - 4:00 p.m. "Ascent of Man" film/discussion, "Majestic Clockwork"
MON. 2 12:00 - 4:00 p.m. "Celebration of the Quilt" exhibition cont.
TUES. 3 10:30 a.m. and 3:30 p.m. Storytime, ages 3, 4, and 5 | 12:00 - 4:00 p.m. "Celebration of the Quilt" cont. | 8:00 p.m. "Careers/Identity" Workshop
WEDS. 4 12:00 - 4:00 p.m. "Celebration of the Quilt" cont.
THUR. 5 10:30 a.m. Westport News Publicity Workshop | 12:00 - 4:00 p.m. "Celebration of the Quilt" | 3:30 Minnybus Matinee [free tickets required] 8:00 p.m. [no tickets] Laurel and Hardy Film Festival: "Your Darn Tootin," "Two Tars," "Leave 'Em Laughing," "Liberty"
FRI. 6 12:00 - 4:00 p.m. "Celebration of the Quilt" cont.
SUN. 8 2:00 - 4:00 p.m. "Ascent of Man" film/discussion, "Drive for Power"
TUES. 10 10:30 a.m. and 3:30 p.m. Storytime for ages 3, 4, and 5
WEDS. 11 1:00 - 2:00 p.m. Valentine crafts for ages 5, 6, and 7
2:30 - 3:30 p.m. Valentine Crafts for ages 8 and up
THUR. 12 LIBRARY CLOSED Lincoln's birthday
FRI. 13 2:00 - 3:30 p.m. "Tarzan the Fearless," film for ages 8 and up [free tickets needed]
SAT. 14 2:00 - 5:00 p.m. "The Denim Expression," workshop on contemporary folk art of handcrafted clothing for ages 12 - 20 [registration required in advance]
SUN. 15 2:00 - 4:00 p.m. "Ascent of Man" film/discussion, "Ladder of Creation"
MON. 16 LIBRARY CLOSED Washington's birthday observed
TUES. 17 10:30 a.m. and 3:30 p.m. Storytime for ages 3, 4, and 5
THUR. 19 3:30 p.m. Minnybus Matinee [free tickets required] 8:00 p.m. [no tickets] | Laurel and Hardy Film Festival: "Sons of the Desert"
SUN. 22 2:00 - 4:00 p.m. "Ascent of Man" film/discussion, "World Within World"
TUES. 24 10:30 a.m. and 3:30 p.m. Storytime for ages 3, 4, and 5
THUR 26 3:00 p.m. Minnybus Matinee [free tickets required] 8:00 p.m. [no tickets] Laurel and Hardy Film Festival: "Come Clean." "Laughing Gravy," "Them Thar Hills," "Hog Wild"
FRI. 27 8:00 p.m. Prism '76 gala opening "Contemporary Westport Painters" exhibition
SAT. 28 2:00 p.m. "The Denim Expression," contest for ages 12 - 20
SUN. 29 "Ascent of Man" film/discussion, "Knowledge or Certainty"

Design • H. Munce

The colored stock, which changed every month, was culled from leftovers that the printer had on hand.

The only way these could have been produced less expensively would have been to use typewriter type for the small stuff. Typewriter type usually has a homemade penny-pinching look. Use it when you must for budget reasons, but set type is preferable.

This is a simple black and white self-mailer. The front cover makes use of a drawing traced from an historical fragment by a young amateur on the library staff.

I choose Benguiat *for the typeface because it smacks of things Egyptian. Actually, it has nothing to do with Egypt—it's just* righter *to my eye than other choices. Having such a feeling for the look of certain faces is one of the satisfactions of dealing with type.*

For the center spread I enlarged the drawing and ran it in a reverse panel.

The drawing sings in this form. It gives the spread a sock and separates the long copy into two relieved sections.

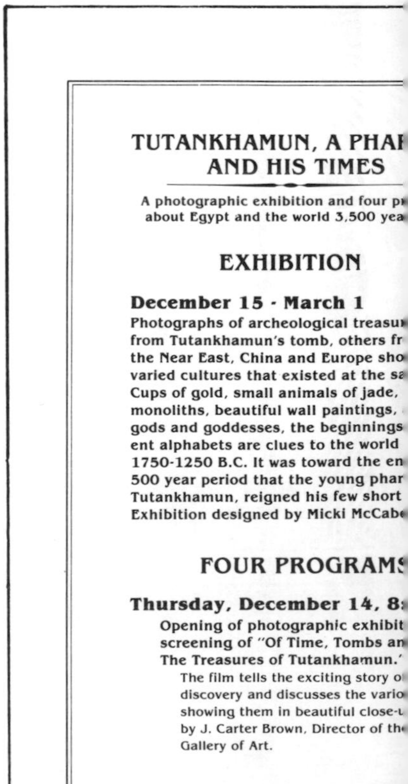
TUTANKHAMUN, A PHA
AND HIS TIMES

A photographic exhibition and four p
about Egypt and the world 3,500 yea

EXHIBITION

December 15 · March 1
Photographs of archeological treasu
from Tutankhamun's tomb, others fr
the Near East, China and Europe sho
varied cultures that existed at the sa
Cups of gold, small animals of jade,
monoliths, beautiful wall paintings,
gods and goddesses, the beginnings
ent alphabets are clues to the world
1750-1250 B.C. It was toward the en
500 year period that the young phar
Tutankhamun, reigned his few short
Exhibition designed by Micki McCab

FOUR PROGRAM

Thursday, December 14, 8
Opening of photographic exhibit
screening of "Of Time, Tombs an
The Treasures of Tutankhamun."
The film tells the exciting story o
discovery and discusses the vario
showing them in beautiful close-
by J. Carter Brown, Director of th
Gallery of Art.

"Beloved of Thoth, the lord of writing"

TUTANKHAMUN, A PHAROAH AND HIS TIMES

A photographic exhibition and four programs about Egypt and the world 3,500 years ago.

at the Westport Public Library in the Meeting Room

"Beloved of Thoth, the lord of writing"

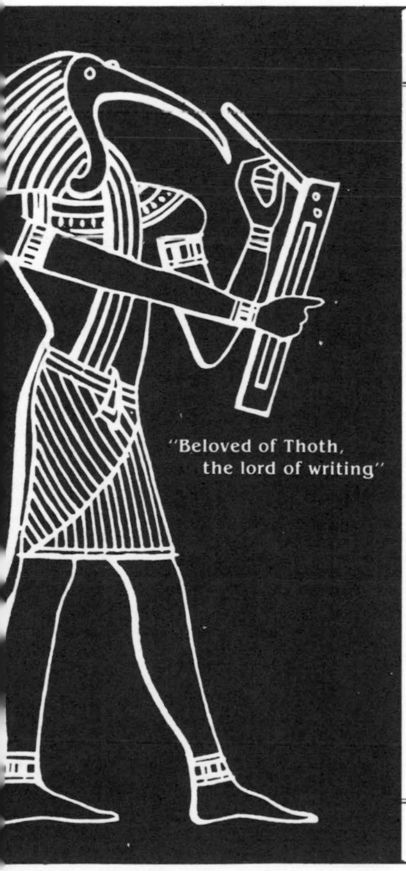

Sunday, January 14, 2:00 p.m.

"Unexpected Egypt: Ancient Egypt in the Light of Recent Discoveries" Slide-lecture by Claireve Grandjouan, Professor of Classics at Hunter College.

Professor Grandjouan will explore fascinating aspects of daily life, religion and art in the incredible civilization to which Tutankhamun belonged.

Sunday, January 28, 2:00 p.m.

"Mysteries of the Great Pyramid" film showing and discussion led by Kenneth Karsten.

The film explains when the pyramids were built and shows their relationship to the time of Tutankhamun. It explores the Egyptians' beliefs about life after death as well as the mystery of how the pyramids were built.

Sunday, February 11, 2:00 p.m.

"Love and Crocodiles: Dramatic Readings from Ancient Egyptian Literature." Introduced by Dr. Karen Polinger Foster, Egyptologist and directed by Albert Pia, Drama Department, Staples High School.

For these dramatic readings which bring to life Egyptian literature, Dr. Foster has selected love poems, stories, and other writings translated from ancient papyri.

Tickets for these free programs may be picked up at the Westport Public Library.

Group visits to the exhibition can be arranged by calling 227-6614.

Stauffer Chemical
Political Contribution Committee

a new democratic process AT WORK.

Activity Report Campaign 1976

CARTOONING
A contest for junior high students will be held at
Westport Public Library
February 10, 1:00 p.m.
RANDALL ENOS, well-known cartoonist - illustrator, will judge the entries.
Bring your own black pens, magic markers, pencils. The library will supply the paper. No application necessary.
AWARDS:
FIRST PRIZE - $50 gift certificate, courtesy of Kleins of Westport.
SECOND PRIZE - $25 gift certificate, courtesy of Max's Art Supplies.
NOTE: Promptly at 1:00 p.m., before the contest, Mr. Enos will give a slide-talk which will be open to everyone — contestants and friends.

This simple, single sheet announcement for the local library allows me to make a couple of points about logic and the organization of elements.

I thought it logical to design the page horizontally so I could set up a strong left to right *movement with the two woodcuts and the flow of the copy.*

The cartoons are not related, but I was amused at the idea of the man diving towards the running sheep—silly, but visually engaging. In addition, those two strong elements in the upper left and the lower right created voids into which the copy could be distributed into organized segments.

Finally I chose Bookman *for the type, a large open, friendly easy to read face. Appropriate, in my view, for the subject of cartooning.*

Photographs of structures as well-known as the Capitol Building, the Eiffel Tower, the Statue of Liberty, etc., exist in abundance—many of them taken so long ago they're in the public domain and you're free to use them.

I started this piece armed with an old photo from the local library's clipping file. First, I had a mezzotint print made, then I enlarged it to make it more grainy.

Next, I let them overlap in the center. This job was printed in red and blue on white stock to further the U.S.A. look. I wasn't sure exactly what effect the blue over red would create where they overlapped, but I was confident it would be interesting.

Artists who are successful in one mode of expression often long to try their hands at another medium. Five illustrators who have managed this change-over and have created interesting works of art as a result are featured in the fifth of the current series of exhibitions at the General Electric Gallery.

The five have moved from canvas to varying forms of three-dimensional construction, using materials that cover the range from clay to wood, metal to plastic. The works have been developed both for gallery showing and as a form of illustration.

Of the five, four are Connecticut residents and one is a New Yorker. All are members of the Society of Illustrators. Each will be represented in a forthcoming Hastings House book on the subject of three-dimensional construction. The book is, incidentally, under the editorship of the GE Gallery's Coordinator, Howard Munce.

MARCH — MAY

GENE SZAFRAN

Born 1941 in Detroit, Michig
and later taught at The Soci
Crafts in Detroit. He has wor
automotive accounts in vari
studios and is now a freelan
His work includes album cov
illustrations, book covers, ch
and annual reports.

Gene's personal interests are
and sculpture, which often b
incorporated into his illustrat
received numerous awards f
and photography, including
Excellence, 17th Annual Exh
Society of Illustrators.

The title of this piece explains the illustration and vice versa.

There are two points to be made here: one is that I luckily happened to own the crude homemade hammer to lend to the photographer—it's much more photogenic than a modern one; and it looks more craftsman-like. Additionally, its proportion to the brushes is exactly right, allowing for the handles to meet at one end and the bristles to emerge past the claw at the other end.

I ran the title in the same manner, from head to handle; this completed the triangular shape.

On the inside spread I gave equal space to each artist for his bio. In cases like this where you must deal with several portraits, try to get the heads about the same size. I wasn't able to do that with Szafran because his picture had no excess area left or right. If I had reduced his head to match the others the photo would have fallen short on the width. Make sure photographers print everything on the negative to insure your options.

NSEL

served U.S. Navy.
ns School of Design, also
as a designer for N.B.C.
rk Times. He now free-
zines and advertising
ork has been cited by the
ub of New York, American
hic Arts, Society of
phis and the Penrose
lpture has been exhibited
of Contemporary Crafts,
His World, Montreal and
ery in N.Y. He is married to
d designer and illustrator,

GERALD McCONNELL

Born in West Orange, N.J. Studied at the Art Students League.

His work is in the collections of the Air Force Museum in Dayton, Patrick Air Force Museum in Florida, the NASA Museum in Cape Kennedy and the National Parks Dept. in Washington, D.C. the Society of Illustrators.

He has served as a trustee of the Frank Reilly School of Art. He was a founding member of the Graphic Artists Guild in 1969 and is currently serving as V.P. of the New York branch.

CAL SACKS

Born — Brookline, Mass. 1914

Studied — Art Students League

He served in the U.S. Navy — in W.W. II as a Graphic artist in the Training Division.

He has paintings in Air Force Art Collection.

Cal works mostly as a book illustrator, and as a designer and letterer for pharmaceutical and general advertising.

He's presently affiliated with American Heritage Publishing Company in a free lance capacity.

WILLIAM S. SHIELDS, JR.

Born in San Francisco — raised in Texas. Attended Chicago Academy of Fine Arts and San Antonio Art Inst. Has worked as a designer/illustrator in Houston, Los Angeles, San Francisco and New York.

Awards include gold medals with New York and L.A. Society of Illustrators' exhibits. Paintings have been exhibited at galleries in Houston, San Francisco, New York and Westport. Is currently engaged in creating assemblages from materials that are, by and large, recycled.

I selected these splendid drawings by Bob Kuhn because they were all done in the same manner and with the same instrument (Nu-pastel). Their rough texture allowed them to be reproduced in line, which gave a look I sought.

Additionally, I picked these particular drawings because the animals were all sitting or lying, and so I could employ a ruled shape to both frame them and give them a "floor line" to rest upon.

This is printed on a textured stock. It's olive green. The ink is an almost black-brown—a rich combination.

Drawings from "The Animal Art of Bob Kuhn" Courtesy North Light and Watson - Guptill Publications

The General Electric Gallery
FAIRFIELD, CONNECTICUT

GENERAL ELECTRIC

General Electric Company, Fairfield, Connecticut 06431

STEVAN DOHANOS, Gallery Director

HOWARD MUNCE, Gallery Coordinator

Feb.-April 1977

"In the case of the true animal artist, picturing animals is not a decision — it is a compulsion which is the dominant force of the artist's entire life and endeavor."

With that as its guiding theme, the Society of Animal Artists was organized in New York City in 1960 by two such artists — Patricia Allen and Guido Borghi. Today the Society's membership includes the finest animal artists in the United States and Canada, some eighty in all.

The Society's President is Paul Bransom, with Anna Hyatt Huntington as Vice President. Both are eminent in their chosen field.

The General Electric Gallery is honored to present a selection of the most celebrated works of the Society's gifted portrayers of the animal world.

SCULPTURE

ROADWAY CONFLUENCE
Hays Van De Bovenkamp

LLANCE
M. Schwebel

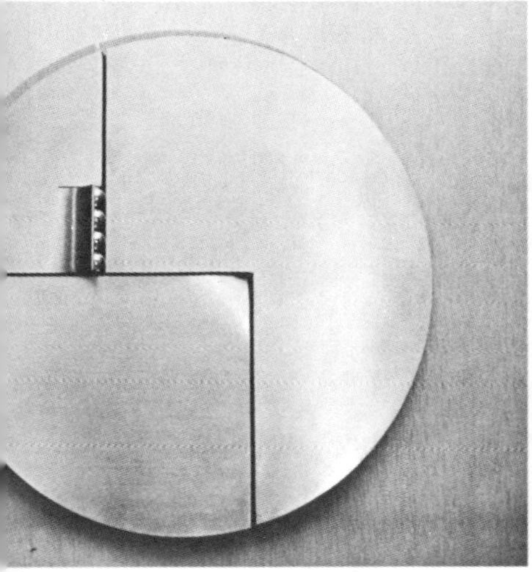

Calvin Albert
Jane Armstrong
Bill Barrett
Helen Beling
Jon Bogle
Isabel Case Borgatta
Rhys Caparn
Leonard Cave
Robert Cook
Carol Kreeger Davidson
Dorothy Dehner
Richard D. Graham
Chaim Gross
Roy Gussow
Cleo Hartwig
Luise Kaish
Nathaniel Kaz
Lloyd Lillie
Bruno Lucchesi
Mashiko
George Mellor
Richard McDermott Miller
Legh Myers
Jene Nickford
Juan Nickford
Abbott Pattison
Marianna Pineda
Arnold Prince
Edward Renouf
Charles Salerno
Renata M. Schwebel
Sidney Simon
Hans Van De Bovenkamp
Ruth Chai Vodicka
Harvey Weiss
Nat Werner
Anita Weschler
Nina Winkel
Jean Woodham

EXHIBITING SCULPTORS

The GE Gallery's first all-sculpture show:

THE MANY TALENTS OF THE SCULPTORS GUILD

For the spectator, viewing sculpture can be a prime experience: 360 times 360 views in one composition, all different, each evolving as part of a harmonic, visual, spatial rhythm. It takes time to make a sculpture—and time to absorb the qualities that are presented.

The General Electric Gallery gives its viewers the opportunity to explore this realm of art by welcoming the Sculptors Guild for its first sculpture exhibit. This non-profit educational organization of professional sculptors has an impressive record of achievement. Founded in 1937, the Guild has presented large annual exhibitions at the Brooklyn Museum, the Architectural League of New York, the American Museum of Natural History, and at the Lever House.

It has assembled numerous traveling exhibits for circulation by the American Federation of Arts, the Carnegie Corporation and the Board of Education of the City of New York, and has also held ten large outdoor exhibitions in New York, one of which transformed Bryant Park into a huge Sculpture Garden, attracting tens of thousands of tourists.

The Guild has sponsored "Competitions for Young Sculptors," and with the help of "Friends of the Sculptors Guild," it hopes to continue this significant contribution to Youth and to the Arts.

November-December 1977

This piece has to be seen and felt to be properly appreciated. It's printed on slick silver stock. The dramatic halftone reproduced beyond my fondest hope. I felt the single word Sculpture *was more dramatic than any wordier headline could possibly be—if the type* looked *sculptural.*

For the inside spread I chose a piece of round sculpture to contrast with the mechanical squareness of all the other elements.

The typeface is Stymie *which comes in several weights, as you can see here. I chose it for its square machined quality—analogous to tools and metals of much contemporary sculpture.*

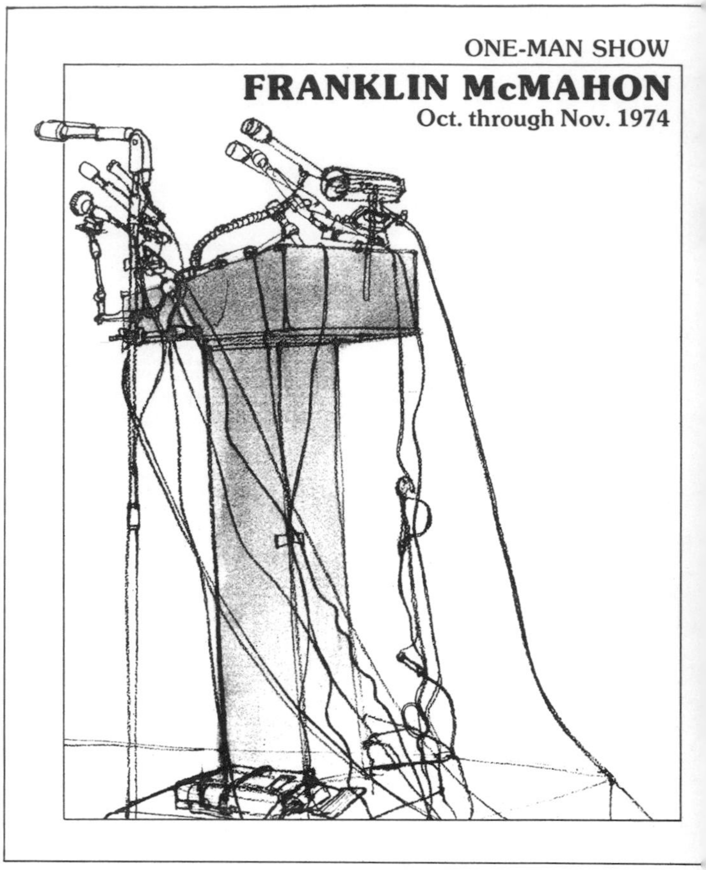
ONE-MAN SHOW
FRANKLIN McMAHON
Oct. through Nov. 1974

This small brochure has two folds and six surfaces, so I was able to be expansive with the elements.

The cover makes use of one of the artist's line and wash drawings, and states its pertinent information in small, quiet terms.

The inside front cover is made from a photo with a dull background. To have included it would have meant reducing McMahon's image; and besides, I wanted a flowing shape instead of a rectangle or square.

So I silhouetted the right side of his figure, left him some objects to lean on, and then let his figure bleed off left and top and bottom.

Be careful when you trim heads: there's a point where figures begin to look mutilated. Remember you're a designer, not an executioner.

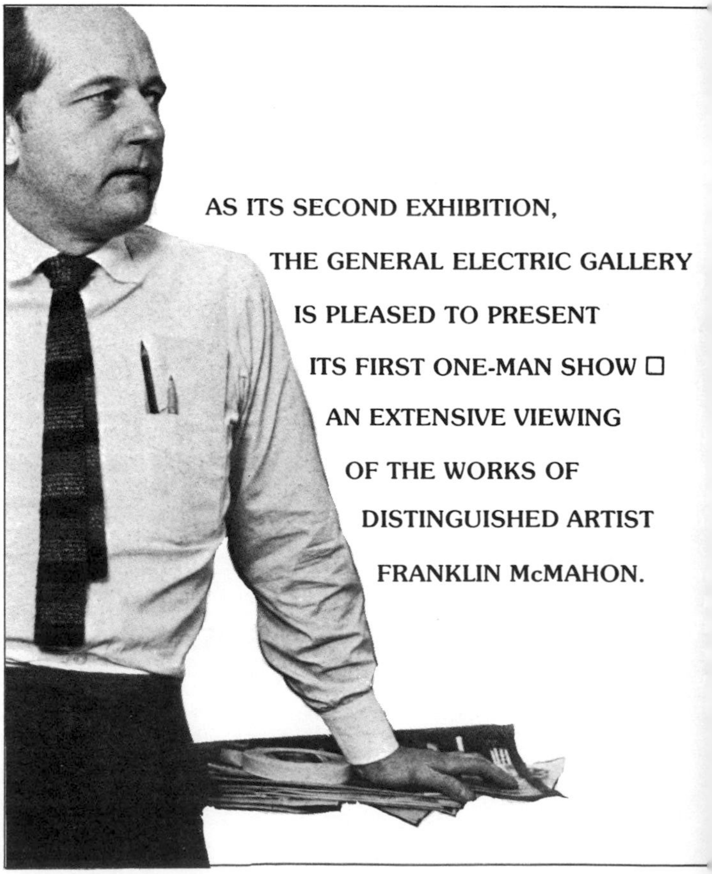
AS ITS SECOND EXHIBITION,
THE GENERAL ELECTRIC GALLERY
IS PLEASED TO PRESENT
ITS FIRST ONE-MAN SHOW □
AN EXTENSIVE VIEWING
OF THE WORKS OF
DISTINGUISHED ARTIST
FRANKLIN McMAHON.

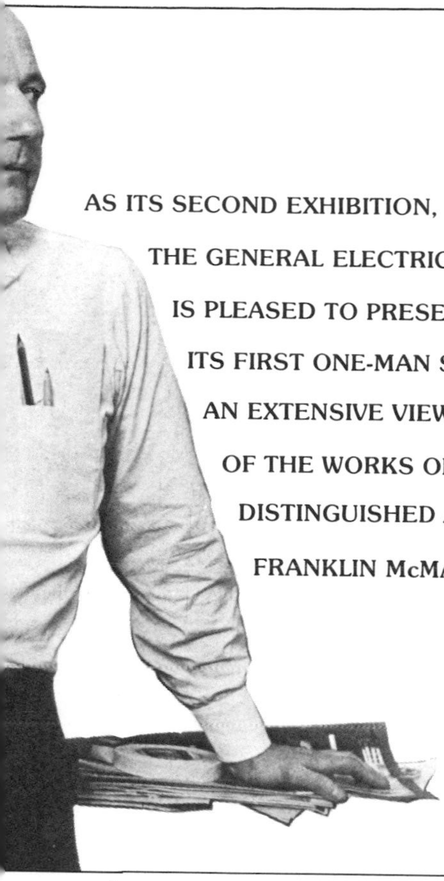

AS ITS SECOND EXHIBITION,
THE GENERAL ELECTRIC GALLERY
IS PLEASED TO PRESENT
ITS FIRST ONE-MAN SHOW □
AN EXTENSIVE VIEWING
OF THE WORKS OF
DISTINGUISHED ARTIST
FRANKLIN McMAHON.

The General Electric Gallery
■ FAIRFIELD, CONNECTICUT ■

This is what you see when you unfold the cover.
. . . And this is what you see when it's completely opened.

McMahon and General Electric meet for a two-
riod when the GE gallery features the works of
inent artist and writer, whose studio is in Lake
inois.
eting, however, is not the first. Earlier,
ı was commissioned to do a painting for the
Electric Calendar. His depiction of work at GE's
chnology Center was featured in the GE
for 1970.
Chicago, McMahon studied at the Art
f Chicago and the Institute of Design. His
and paintings have been featured in *Fortune,*
ustrated, The Saturday Evening Post, McCall's,
uesday and World Book Year Book.
orts illustrator he has covered goose hunting in
e San Diego-Acapulco yacht race, partridge
Spain, activities of the Royal Bangkok Sports
the Kefauver hearings into professional boxing.
active theme for this versatile artist has been
e of the U.S. space program. He was on hand
nching of Gemini III, aboard the USS Wasp
for the recovery of Gemini IV, at the Canary Island tracking station for Gemini VI, at Cape Kennedy for Apollo X and at Mission Control in Houston for Apollo XI, when man first walked on the moon.

He has also taken the political trail. Following the candidates through the primary elections, the conventions and the campaigns, he produced works that were published by leading newspapers as well as by *Fortune* magazine. But in addition his drawings were developed into an hour-long documentary, *Portrait of an Election,* shown on CBS. It was awarded an "Emmy" by the Chicago Chapter of the National Film Festival.

His paintings have formed the basis for other films, including *World City,* which was based on his trips to Europe, the Far East, Middle East, and South America.

In 1967, Franklin McMahon was named Artist of the Year by the Artists' Guild of New York. He also holds the renaissance Prize of the Art Institute of Chicago and is an honorary member of the Artists' Guild of Chicago and Chicago's Society of Typographic Arts.

Conover
Gallagher
O'Connor

Investigative & Security Consultants, Inc.

This assignment called for a brochure announce the launching of a new investigative service headed by three ex-F.B.I. men.

Among the other attributes they broug to their new firm were their three euphonious Irish names. When I wrot the names out and examined them, I pleased to see that the first letter in ec case was a round one. Good start.

Next, I perused my type books lookin C's, G's and O's that had distinctive characteristics and went well together found these three in a photo alphabet called "Torino Flare."

Then I chose Baskerville Bold *for th maining letters. It was necessary to hc this degree of weight and blackness to wed with the capital letters.*

This piece gives me an opportunity to make a small speech about not being tent with design results that don't wo entirely to your satisfaction and whim For instance, in this piece I would ha liked to have put the remaining letter Gallagher into the open slot of the G. looked good—and it landed in the cer of the other two lines—the only probl was that it didn't read fast enough ar broke the rhythm of the eye.

When it came to placing Connor, I tr it in all positions from the top to the tom of the o, and stopped where it loc best. That's the simple answer to ma decisions in design—the rightness of something makes you feel good.

We'll deal with other logotypes in the next chapter.

The Levitational Realism of Gary Erbe
February–April 1979

This brochure introduced the work of Gary Erbe, who called many of his paintings "Levitational Realism"— objects suspended in air.

So I took him at his word—or words: I suspended the picture of his painting up in the air, too. Then, to add to the effect, I ran the title in small type at the bottom and used an old-fashioned dingbat to point back up. I repeated the pointing finger on the inside spread and levitated the gallery logo and credits on the back cover.

Always examine the words and copy you're dealing with to spark graphic ideas.

Levitational Realism

It's the original, provocative style developed by painter Gary Erbe.

Gary Erbe's paintings are in the trompe l'oeil tradition but transcend it to achieve a meticulous and imaginative style all his own. He calls it "Levitational Realism." His paintings are scrupulous juxtapositions of carefully selected objects freed from their natural environments via levitation.

Located in space, strongly lighted, the objects of his paintings radiate a timelessness. Erbe's stated objective is to intimate, through specific objects portrayed in highly-modeled, fool-the-eye assemblages, such universals as the pain of loss, the inevitability of change and the persistence of hope.

Born in Union City, New Jersey, self-trained Erbe began painting in 1965 and exhibiting his work in 1970. He has participated in shows at the New Britain Museum of American Art in Connecticut; the National Academy of Design, the Salmagundi Club and Allied Artists of America in New York City; the New Jersey State Museum, Newark Museum, Jersey City Museum, and at several New Jersey colleges and one-man shows.

Mr. Erbe is a member of the Allied Artists of America, Salmagundi Club, and Associated Artists of New Jersey.

Photo, C. R. Shelton

DESIGNING A LOGOTYPE:

Otherwise known as trademarks, slugs or corporate identifications. Try it. You'll like it.

Designing logotypes or "slugs" as they are sometimes called is a very pleasurable pursuit.

It's a design problem in which luck plays a large part. You're in luck if the company or organization you're doing it for has a happy combination of letters in their name, or if they're in an unusual business that allows you to use fresh and unhackneyed symbols. The reverse of that would be to have to do a logo for anything to do with airlines or aeronautics. There can't be many new ways left to portray wings and flight.

A good logo demands the following qualities:

1 *It must identify the organization when it stands alone.*

2 *Its look should reflect the tone of the organization.*

3 *It must be simple enough to withstand severe reduction.*

4 *It must be adaptable enough to serve such opposite purposes as a letterhead or a marking on a crate or a TV spot.*

Good logo design is an elusive thing. You can find many that appeal at first sight, but because they're a bit too slick and tricky, don't wear well. Many are too trendy—merely designer's conceits.

Compare some of those with established classics such as General Electric, Ford, Prudential, CBS, Kellogg's or Coca-Cola. To be sure, these latter all have had long exposure and millions of dollars spent on them—but they're all basically good and have distinct personalities.

Many of them were designed and established long ago when they were simply called trade marks. Their managements wisely held on to the investment they had in them with the public. Instead of scrapping them as old-fashioned and corny, as many foolish corporations have done with theirs, they cleaned them up, simplified, and modernized them over a period of time without the public being aware of the change.

Some that were probably done by sign painters originally have been ingeniously transformed into very contemporary-looking pieces without anyone becoming the wiser.

Serif Gothic Outline #2

AABCDEEFGHIJKLLMMNNOPQRS
TUVVWWWXYZ&.,:;!?'""/[]()()*·--

Here is an example of what I meant by luck playing a part in logo design.

This one of mine was designed for General Electric's 100th Anniversary Year.

I approached the problem as I do with all logo assignments; first I write out the name of the organization, then reduce it to its initials and whatever other information is required. In this case, the numerals 100 were as important as anything else.

Next, I went through all my type and phototype books and examined every alphabet for G's, E's, 1's and zeros. After much discouragement, I came upon what proved a real find. In an alphabet called "Serif Gothic Outline," which gives you a choice of a square capital E or a round one, the center bar of the latter fortunately is in the same position as the cross piece of the cap G. So by butting the two letters together it was possible to make the center bar serve for both letters, thereby forging them into a single unit.

My luck held still further! Since the G and the E were both based on a perfect circle it was then possible to join the two zeros to each other and then to the two letters and the numeral 1. The result was a distinctive logotype which, because it dealt with a celebration year, was used on every conceivable kind of gift from neckties to jewelry to flags to plaques.

I never had a more satisfying exploitation and exposure of my work.

sundays child

1165 Park Avenue • New York, N.Y. 10028 • (212) 575-0704

Two happy S's—one on each end—they have a kind of comical look which was appropriate for this kid's newspaper supplement. I did the word "child" as handwriting to keep it light and informal. Dealing with children's subjects is touchy territory. Be on guard against getting too cute. . . like showing letters backwards and other embarrassments.

the Suburbanite

This store had been using an old logo based on a coat hanger when I got this assignment. I suggested they hold on to the idea, but to update it. They agreed. So I used this beefier type of hanger to replace the spindly original, and used their name as the cross bar instead of having to put it elsewhere. The heavier version also lends itself better to woven labels. . . important to a haberdasher. Sometimes it's used within an oval and sometimes by itself.

the Suburbanite

Dedicated to pleasing those men who want clothes lasting in style and enduring in material.

Route 7 Opposite the Wilton Cinema
Open Thursday Evenings

the Suburbanite

Dedicated to pleasing those men who want clothes lasting in style and enduring in material.

Route 7 Opposite the Wilton Cinema
Open Thursday Evenings

The hangers sometimes serve as decorative brackets.

Suggestions to bring joy to a man at Christmas

AND LADIES! Southwick blazer for you **180.**

1 Tweed jacket by Southwick **175.**
2 Wool Argyle socks **3.00.**
3 Pure Camel Hair Polo Coat **320.**
4 Corbin prime flannels **52.50.**
5 Plaid Viyella sport shirt **42.50.**
6 Southwick suit **220 to 265.**
7 Ready tied silk Ascot **12.50.**
8 Sheepskin shearling coat **330.**
9 Checked wool slacks by Corbin **55.00.**
10 Woolen sport vest **27.50.**
11 Velour shirt **36.50.**
12 Scottish shetland sweater **31.50.**
13 Bladen of England sport coat **295.**
14 Begg of Ayr wool scarf **14.50.**
15 Norman Hilton suit, from **295.**
16 Alpaca wool cardigan **41.50.**
17 Cashmere turtleneck dickey **32.00.**
18 Real shearling gloves **45.00**
19 394 Neckties **12.50 to 25.00.**
20 Pendleton wool shirt **29 to 32.50.**
21 Cotton chamois shirt, Woolrich **21.50.**
22 Made-to-order suit **235 up.**
23 Lyle turtleneck shirt **14.**
24 Leathersmith pocket diary **10.**
25 "British Warm" overcoat **210.**
26 Lined dress gloves **24.**
27 Mighty Mac pile lined jacket **100.**
28 Cavanagh-edge felt hat **35.**
29 Corduroy sport coat **115.**
30 Fancy blazer buttons **10.**
31 Moleskin slacks **50.**
32 Leather wallets **13.50 to 50.00.**
33 Dopp kit, cowhide **16.50.**
34 Navy blazer **125-155.**
35 Leather eyeglass holder **12.50.**
36 London Fog lined raincoat **90.**
37 Fleece lined mittens **11.50.**
38 Circingle belt **12.**
39 Wool knit watch cap **8.**
40 Silk bow ties **8.50.**
41 Flannel-lined chinos **29.50.**
42 Wool Gabardine Raincoat **175.**
43 Folding umbrella **17.**
44 Alan Paine lambswool V-necks **42.50.**
45 Viyella bathrobe **77.50.**
46 Izod alligator cardigan **25.50.**
47 Marino wool turtleneck **35.**
48 Woolrich shirt jacket **36.50.**
49 Down vest **65.**
50 Classic button down shirt **19.50.**
51 True hunting jacket, down liner **155.**
52 Quality corduroy slacks **35.50.**
53 Cashmere sweater **85.**
54 100% camel hair blazer **220.**
55 Royall Lyme lotion **10.**
56 Pleated trousers, **41.50 up.**
57 Sueded chamois shirt-jacket **155.**
58 Tweed cap **15.**

Open Thursday Evening until 9.

the Suburbanite

17 Danbury Road (Route 7) Wilton   Phone 762-5177

For Christmas a sprig of holly makes it seasonal.

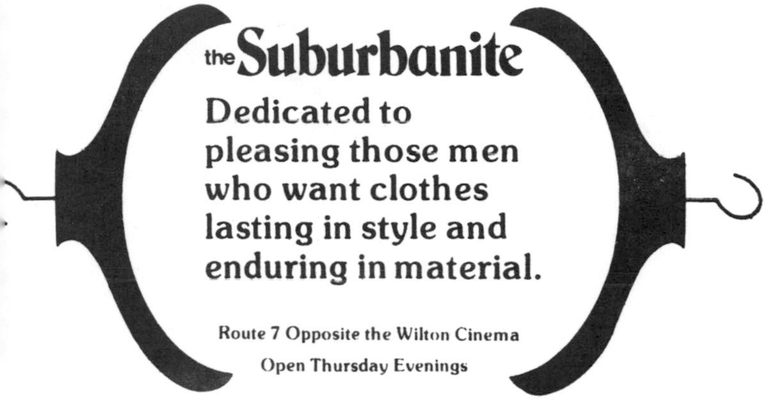

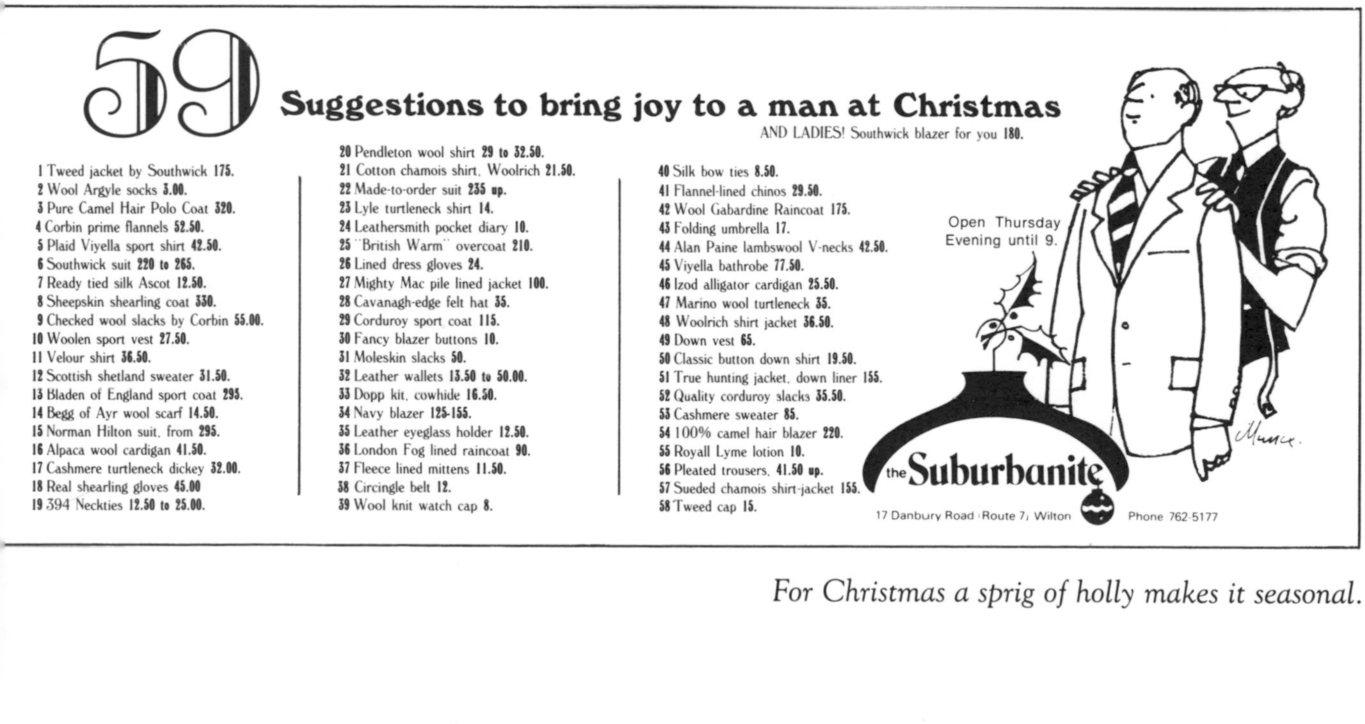

1 AMERICAN RED CROSS

2 ALCOHOLISM GUIDANCE CENTER

5 COUNSELING SERVICE (WESTPORT-WESTON)

6 CEREBRAL PALSY

9 FAMILY & CHILDREN'S AID

10 GIRL SCOUTS

13 OPEN LINE

14 RENAISSANCE

17 U. S. O.

18 VISITING HOMEMAKERS

This was for a two-color pamphlet needed in a hurry. Each of the organizations needed to be identified by a logo—but none existed in several cases. So I hastily put together some temporary ones out of my type files, using only initials. They are numbers: 4, 5, 7, 8, 11, 13 and 18.

3 BOY SCOUTS

4 CARVER FOUNDATION

7 CHILD GUIDANCE CENTER

8 COMMUNITY COUNCIL

11 INTER-COMMUNITY CAMP

12 MENTAL HEALTH ASSOCIATION

15 SALVATION ARMY

16 SOCIETY TO ADVANCE THE RETARDED

19 WESTON LITTLE LEAGUE

20 Y. M. C. A.

North Light
Book Club

P.O. Box 489 • Westport, Conn. 06881 • 203/226-3533

The words North Light and Book Club are almost identical in length. So I ran the first two words left and the other two right (thereby creating the voids for the pencil and brush to fit into). I framed each in a rectangle to achieve the sharp black corners which help to make another rectangle of the whole unit. Note how it helps achieve cohesiveness when you can (or are lucky enough to) tuck letters together as happens here with the descender of the g *and the capital* B.

THRUSH PRESS, INC.
52 E. 19 ST.
N.Y.C. 3, N.Y.

Anyone connected with the communications and printing business knows that deadlines and clock-fighting never cease.

No matter how early projects are begun, Time, precious Time, evaporates along the way and things wind up in a rush.

The word "Rush" is the key to this simple design I did for a printer friend with the intriguing family name of Thrush.

It's one of these happy circumstances where an idea lurks in a fortunate find like this, or in a happy combination of letters. A design of a winged thrush might have been a possibility here—but too cute. I prefer this tough, meaningful command.

Train yourself to look for such finds in all your visual problems.

HANSON COMMUNICATIONS, INC., P.O. BOX 1350, NORWALK, CONNECTICUT 06852 (203) 838-5566

wnlk 1350

FEATURES

ORWALK ... WESTPORT

WILTON ... FAIRFIELD

EW CANAAN ... WESTON

The problem here was to design a press sheet that would encompass the names of the eight communities that this radio station served.

These sheets comprised a color-coded kit—a different color for each subject.

My solution was to start with a solid, gutsy typeface (Memphis) that has a square serif which would in effect create a visual line along the bottom or top of each word.

Then I ran the type in its negative state (white) and printed a light tint (but strong enough to define the letters) behind the type.

Since the negative words were really the white paper, surrounded by tint, the bottoms of each letter melded into the white of the center of the sheet and formed a sharp rectangle.

This letter form, at this size and weight made it possible to say what needed to be said and to do it decoratively at the same time. (I was also lucky to have eight names to deal with—two at each side and top and bottom.)

Be grateful for luck—it often makes a designer look good.

The word Film *says "entertainment" to me; not always jolly, to be sure, but often enough to chance a few frills—hence the choice of this fancy letter form for a logotype to be employed for many different functions.*

The circular version is a label for the containers that film is transported in. It fits in a depression stamped into the can and appears to advantage against its dark metal background. The film reel that dots the "I" adds to the circular motif. This is in two colors.

It was used in black and white for stationery. Even though the reel is cropped, there's enough shown to serve its purpose.

The word "movies" certainly has an air of fun about it—so I employed the fat and friendly features of Cooper Black *to convey that mood.*

Westport
Public
Library

FILMS

hursday night movies

Thursday Night Movies at the Westport Public Library will be shown every week through the end of May. These adult film programs are free and everyone is welcome. Announced titles are subject to occasional change.

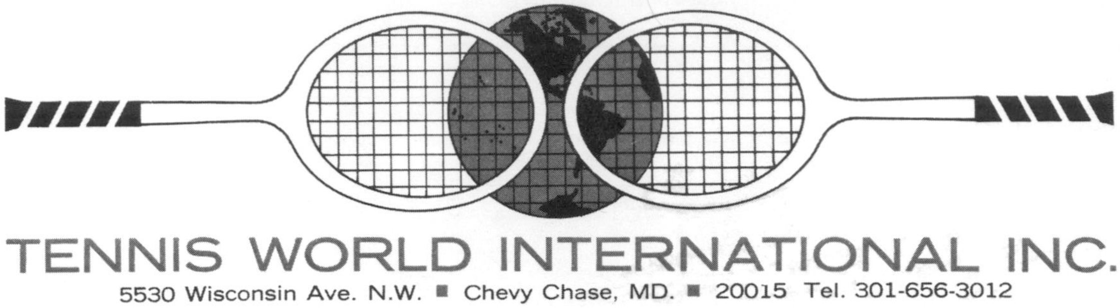

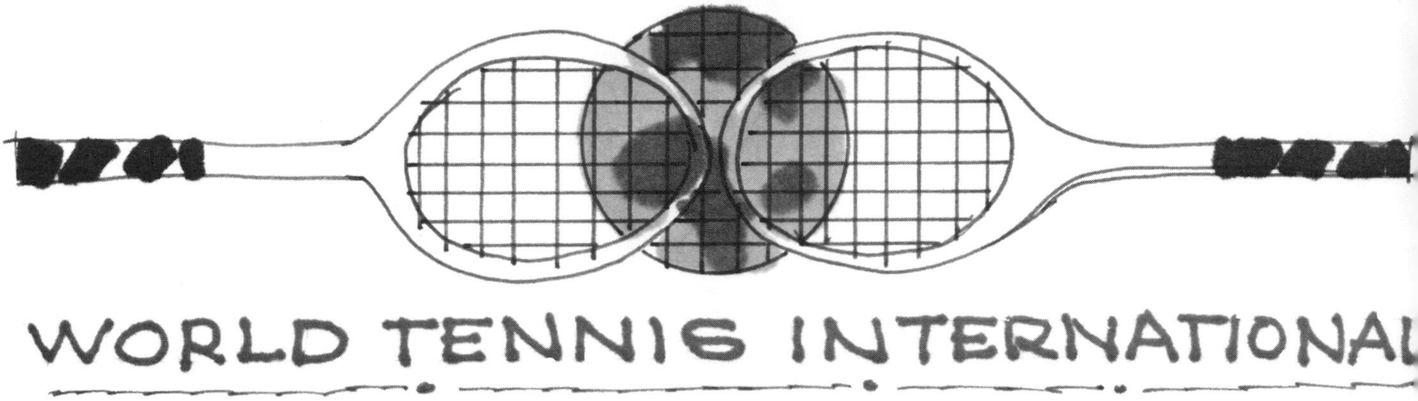

The globe is one of the most often used symbols in logo design; every company that does international business likes it to be known and shown. That's obviously how it was with this assignment, which not only said "International" but "World" as well.

Again, I was fortunate in that it had to do with tennis—and the strings of the racket automatically created the grid lines of the globe and vise versa.

This was a two-color job—often not possible in logo design. However, if I had been restricted to black only I would have screened the black to make the ocean areas gray instead of the green as they were used.

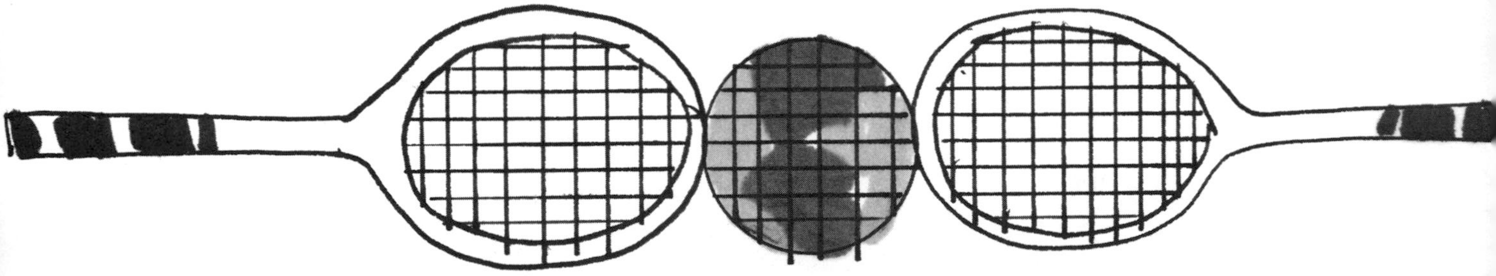

This is less effective than the chosen one which portrays the same thing but doesn't let the strings double as grids.

Here are a few of the other rough idea sketches I submitted.

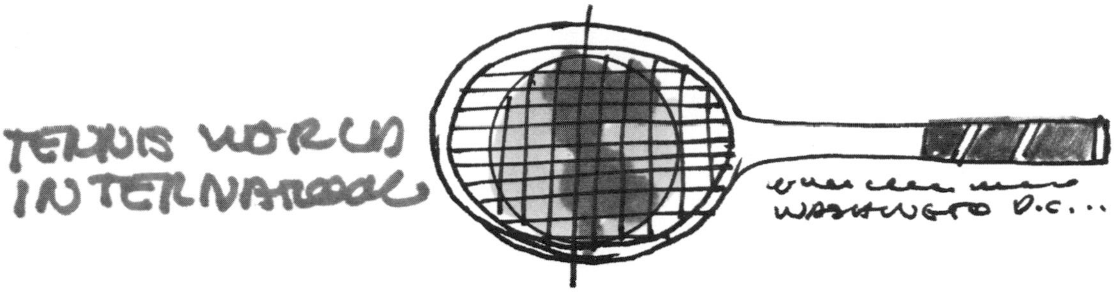

I liked this one too. It had the added attribute of performing the whole job with one racket.

This began to look like a face wearing sunglasses—no good.

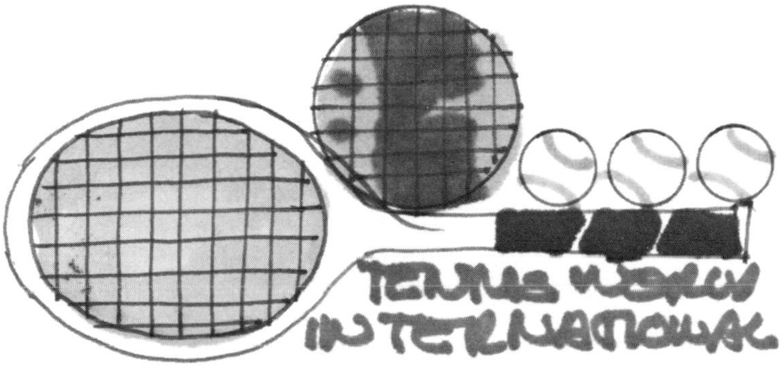

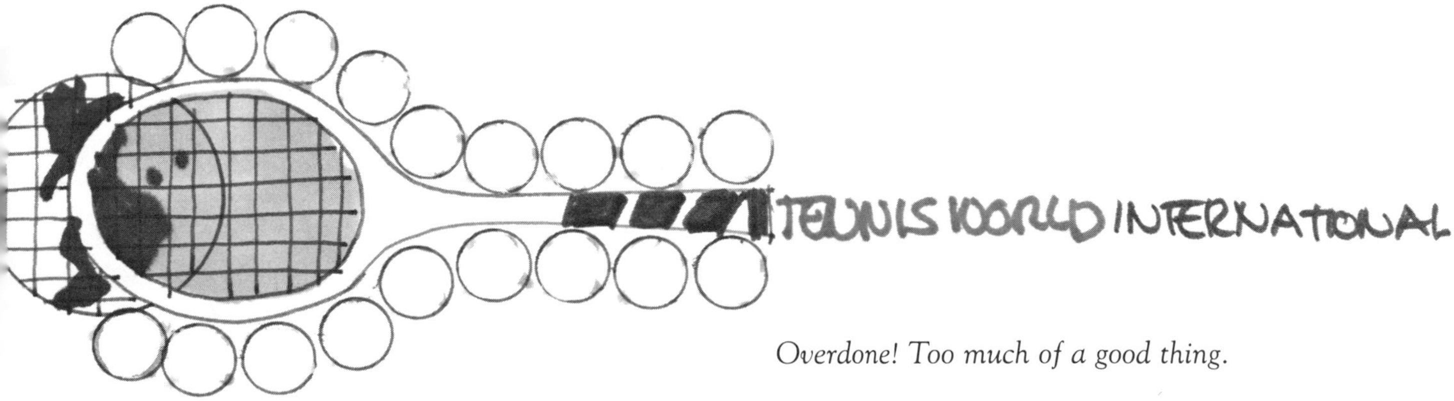

Overdone! Too much of a good thing.

The real estate pages of newspapers these days contain many fresh and inventive logotypes—often combining simple stylized art with distinctive lettering or type. Here are several examples for condominiums and resorts.

Saul Bass

Here are two excellent logos—a national one and a local one.

You'll recognize the top one as belonging to the Girl Scouts; this revised version retains the original overall shape, and ingeniously imbeds the heads of three females, black and white, within it.

The lower one speaks for itself. Max's is, of course, an art supply store. It was designed by James Williams of Easton, Conn.

55. Colorforms Inc., Norwood, New Jersey (1960)
Designer: Paul Rand

56. Columbia Records, New York, N.Y. (1954)
Designer: S. Neil Fujita

139. Oroweat Baking Company, Los Angeles, Calif. (1963)
Designer: Richard Runyon Design

140. Pepperidge Farm, Inc., Norwalk, Conn. (1956)
Designer: J. Roy Parcels/Dixon & Parcels Associates, Inc.

149. Push Pin Studios, New York, N.Y. (1954)
Designer: Push Pin Studios

150. Jorge R., Los Angeles, Calif. (1963)
Designer: Douglas Carr/Tilds & Cantz Advertising

21. American Broadcasting Co., New York, N.Y. (1962)
Designer: G. Dean Smith

22. American Cast Iron Pipe Company, Birmingham, Ala. (1960)
Designer: Robert Luckie & Co., Inc.

159. Rural Mutual Insurance Co., Madison, Wis. (1959)
Designer: Thomas Laufer & Associates

160. Scott Paper Company, Philadelphia, Pa. (1962)
Designer: Hiram Ash Inc.

63. Crown Zellerbach Corp., San Francisco, Calif. (1952)
Designer: Frank Gianninoto & Associates

64. Culligan Inc., Northbrook, Ill. (1961)
Designer: Charles MacMurray/Latham, Tyler, Jensen, Inc.

161. Seattle Copy Club, Seattle, Wash. (1959)
Designer: Robert R. Overby

162. Seiberling Tire & Rubber Co., Barberton, Ohio (1962)
Designer: Scherr & McDermott International, Wyant Bldg,

65. Dayton Typographic Service, Dayton, Ohio (1963)
Designer: Hahn & Costello Inc.

66. Design Built Exhibits, Inc., Long Island, N.Y. (1962)
Designer: George Tscherny

147. Pounder's Poultry, Akron, Ohio (1963)
Designer: Richard Oborne

148. Powow In The Pines, Inc., Amesbury, Mass. (1963)
Designer: George E. Hands/Center for Industrial Design

19. Alpine State Bank, Rockford, Ill. (1962)
Designer: Howard H. Monk Associates

20. The Aluminum Association, New York, N.Y. (1963)
Designer: George Tscherny

Here are several examples of excellent professionally designed logotypes. Study them. Discover for yourself what makes them tick.

67. Diamond Crystal Salt Co., St. Clair, Michigan (1961)
Designer: Dickens, Inc.

68. A. B. Dick Company, Chicago, Illinois (1963)
Designer: Walter Dorwin Teague Associates

43. Books, Unlimited, Sarasota, Fla. (1962)
Designer: Robert Wesley

44. Bow Hunter Club, Fayetteville, N. Carolina (1960)
Designer: Harry Sehring

69. Ditto, Inc., Lincolnwood, Ill. (1961)
Designer: Morton Goldsholl Design Associates

70. Diversified Products Corp., Opelika, Ala. (1963)
Designer: Frank Cheatham/Porter & Goodman Design Associates

45. Brentwood Shopping Center, New York, N.Y. (1958)
Designer: Dominic Arbitrio/Page, Arbitrio & Resen

46. Brown-Forman Distillers, Louisville, Kentucky (1958)
Designer: Raymond Loewy/William Snaith, Inc.

95. Hartford Public Library, Hartford, Conn. (1961)
Designer: William Wondriska

96. Helbros Watch Co., New York, N.Y. (1948)
Designer: Paul Rand

31. Bachman Bakeries Corp., Reading, Pa. (1957)
Designer: J. Roy Parcels/Dixon & Parcels Associates, Inc.

32. Bank of Texas, Houston, Texas (1963)
Designer: George K. Buckow, Jr.

97. Helipot Div. of Beckman Instruments, Fullerton, Calif. (1948)
Designer: Advertising Designers Inc./Lou Frimkess

98. Hill Top House, Dayton, Ohio (1963)
Designer: David Battle/Vie Design Studios, Inc.

33. Barr Chemical Inc., Berkeley, Calif. (1962)
Designer: G. Dean Smith

34. Bauer & Black, Chicago, Ill. (1961)
Designer: Morton Goldsholl Design Associates

107. International Farmers Market, San Mateo, Calif. (1961)
Designer: Gollin & Bright, Inc.

108. International Harvester, Chicago, Ill. (1946)
Designer: Raymond Loewy/Wm. Snaith, Inc.

119. Marine Sciences Corp., Gardena, Calif. (1963)
Designer: Frank Cheatham/Porter & Goodman Design

120. Nancy Lee Martin, Los Angeles, Calif. (1962)
Designer: Jerry Braude

109. International Telephone & Telegraph Corp., N.Y., N.Y. (1960)
Designer: Matthew Leibowitz

110. Japan Airlines, San Francisco, Calif. (1960)
Designer: Bruce Montgomery/Robertson-Montgomery

121. Martin-Senour Company, Chicago, Ill. (1954)
Designer: Morton Goldsholl Design Associates

122. Menasha Corp., Menasha, Wis. (1962)
Designer: Hudson-Wolter & Associates

ART WORK:

Several suggestions for the acquisition of inexpensive art to decorate & enhance your printed piece.

There are many occasions when it's desirable or necessary that a printed piece be done all-type with no illustrations, no photos and no decoration. But your life would get pretty dull if you were permanently assigned such a fate.

Let's assume that drawing is not one of your strong points and you don't have the money to retain an illustrator or cartoonist, but you want very much to illustrate or somehow visually enhance your printed message—to give spark to the gray mass of type and, most important, to make the piece *inviting* to look at and to encourage the viewer to read on.

Here are some options:

If you are volunteering your efforts for a charity or church or civic project—and no money is changing hands—see if you can find in your group someone as generous with their time as yourself who can draw well and will work with you.

Failing this, check with the art teachers in your local school to see if they'll introduce you to some talented students who will undoubtedly be thrilled to have their work reproduced for the first time.

If you're lucky, you may find one of their existing drawings that's appropriate. If not, see if you might assign them your problem. Be sure to give them a credit line or run their signature.

Your next possibility is to use some form of stock art. There are two kinds: one contemporary and one with an antique flavor. There are catalogs of art you can buy with dozens of illustrations in many styles to choose from. They are rendered in decorative, cartoon and realistic manners.

They are usually in the form of line art and therefore present no problem or extra cost for reproduction.

The other form of stock art—mostly quaint drawings and steel engravings are the product of an earlier advertising age. These have been collected in books to be found at your library. They are in the public domain, so there's no cost when you reproduce them.

Whether it's the library's copy or your own, photostat or xerox the drawing you wish. I suggest you get on the mailing list of Dover Publishers, 180 Varick St., New York City, NY 10014. They specialize in such collections plus books on symbols, old typefaces and a myriad of other specialties that you'll find helpful. These books are inexpensive and excellent. They're constantly used by professionals.

In addition to art work that depicts people and things, there's another form to be aware of: decorative type borders. Every type house has a collection and they're also now available in transfer sheets. Both sources can also supply you with arrows, corners and Spencerian curliques of every description.

And here's a bit of incidental intelligence—Archie Bunker's pet name for his late wife Edith, "Dingbat," is a printer's term and it means those little odd art-spots that were once used to separate printed items or to take up the slack when a column fell short.

The best of these are very good indeed—they're often funny or otherwise attractive. When you're lucky, you can find one absolutely germane to your assignment. Enlarge it, and you have a needed illustration.

Photographs are illustrations, too, of course, but we'll deal with them separately—in the next chapter.

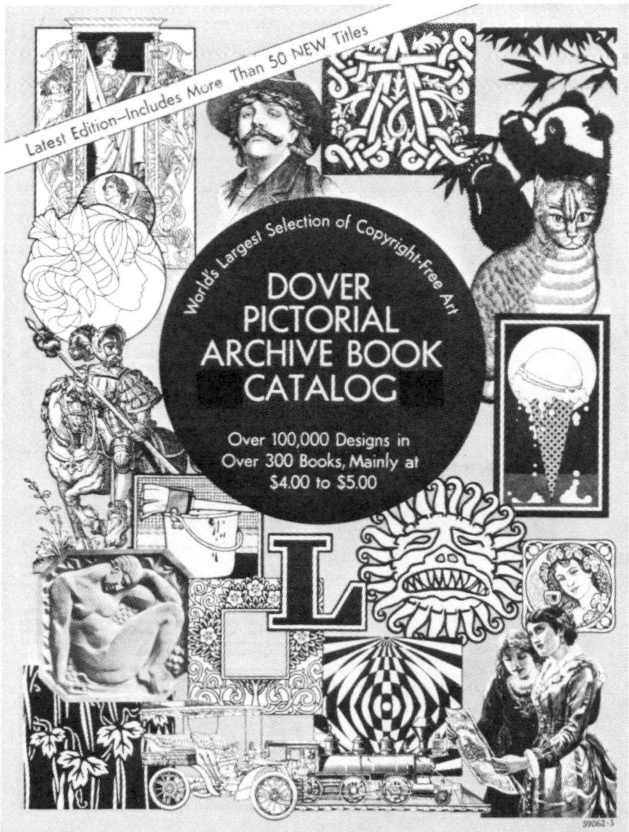

Pictured here and on several subsequent pages is some of the ready-made art work available to you from Dover Publications.

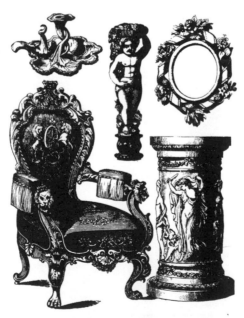

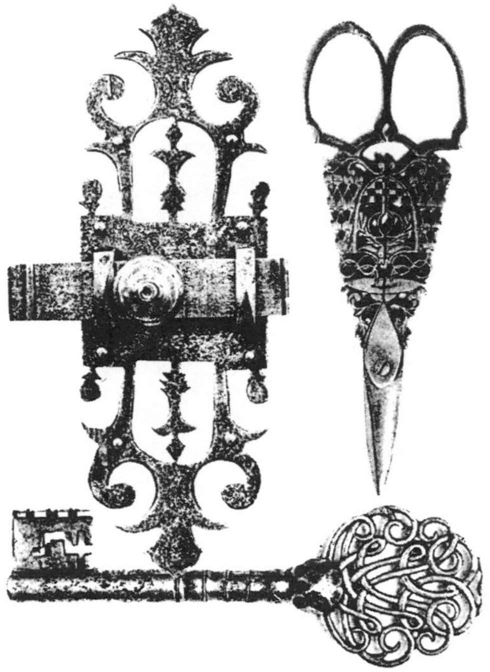

DECORATIVE ANTIQUE IRONWORK
Henry R. d'Allemagne

Photographs of 4500 iron artifacts from world's finest collection, in Rouen. Hinges, locks, candelabra, jewelry, weapons, lighting devices, clocks, tools, from Roman times to mid-19th century. Indispensable for artist and craftsman. Nothing comparable to it. 420pp. 9 x 12.

HANDBOOK OF ORNAMENT
Franz S. Meyer

Classic collection, used by generations of artists, designers, illustrators, students; 3000 line cuts of decoration, decorated objects: Classical, Medieval, Islamic, general European, Grotesques, architectural features, chairs, cabinets, crowns, lanterns, weapons, bells, pillars, altars, lamps, vases, etc. Equal to a whole file cabinet of pictures. 548pp. 5⅜ x 8.

THE STYLES OF ORNAMENT
Alexander Speltz

Complements Meyer's Handbook. Encyclopedia of design, application of ornament. 3750 illustrations from primitives, Egyptians up to Victorians cover ceramics, jewelry, lighting devices, furniture, windows, musical instruments, clothing, coins, too much else to list. One of the 2 standard handbooks. For redrawing, direct copying, identification, decoupage, toy projects. 647pp. 5⅝ x 8⅜.

HISTORIC ORNAMENT, A PICTORIAL ARCHIVE
C. B. Griesbach

900 representations of ornamented objects, ornament, Ancient Egypt to 1800: Greek vases, Classical orders, Byzantine mosaics, Romanesque and Gothic architectural features, textiles, weapons, bookbindings, fountains, patterns, geometrics, woodcarving. Finer quality than most historical collections. Reference, ideas, direct reproduction. 280pp. 8⅜ x 11¼.

350 large illustrations clearly reproduced

FOOD AND DRINK: A Pictorial Archive From Nineteenth-Century Sources

Jim Harter

This monumental collection contains over 350 illustrations of every conceivable activity concerned with the preparation, display and consumption of food and drink. Commercial designer and collagist Jim Harter combed many rare 19th-century graphic sources to find this unusual, eye-catching, mouth-watering material. From the dining car of the Orient Express to the family kitchen, exotic and familiar cuisine is presented in unmatchable line drawings sharply reproduced. Pictured are rajahs at table in their palaces, cavemen squatting over their catch, Romans feasting, tramps on bread lines, elegant restaurants, public kitchens, Christmas dinners, court banquets, coo tasting, kids eating, the full ritual of food. Of course individ edibles are represented, including cakes and sweets, fish, melo large and small, grapes, artichokes, rhubarb, leeks, pumpkins, pi poultry, etc.; and the instruments of food are not neglected, w glassware, tea sets, decanters, mugs, pitchers, baskets, bowls, ur ladels, cutlery all shown, as well as menus, wine lists, etc. A tr cornucopia of plenty for collagists, decoupeurs, any artist needing readily accessible food motif. 144pp. 8⅞ x 11¾.

WOMEN: A Pictorial Archive From Nineteenth-Century Sources

Jim Harter

Women of all kinds, types, shapes, ages, occupations, all in sharp black and white for best reproduction. 391 illustrations from rare 19th-century periodicals show Grecian maids, Victorian maidens, dancers, housewives, courtesans, goddesses, fairy princesses, women working, playing, laughing, crying, flirting. 120pp. 9 x 12.

CHILDREN: A Pictorial Archive From Nineteenth-Century Sources

Carol Belanger Grafton

242 rare wood-engraved illustrations specially selected for quality draftsmanship from *Chatterbox, British Workwoman* periodicals, p ture children being and doing. Rich boys, beggars, sporting typ school kids, pretty little girls and babes, all shown in detailed per settings or background free for easiest use. 119pp. 8⅜ x 11¼.

SILHOUETTES: A Pictorial Archive of Varied Illustrations

Carol Belanger Grafton

This collection of over 600 silhouettes from the 18th to the 20th centuries presents a most unusual and useful archive. Few illustrations can be more starkly dramatic or atmospheric than a silhouette; yet silhouettes (to say nothing of *good* silhouettes) are more scattered than most kinds of graphic art. Here in an inexpensive volume are not only portrait and profile silhouettes, but groups and scenes, an alphabet, birds and animals, trees and other still life, beautiful ships. There are men, women and children in full figure and profile, engaged in many activities, dressed and styled in period elegance from rococo to Art Nouveau. Unsurpassable for spot illustration, needlework design, greeting cards. 144pp. 8⅜ x 11¼.

MEN: A Pictorial Archive From Nineteenth-Century Sources

Jim Harter

Nearly 400 steel engravings from such periodicals as *Leslie's, Illustrated London News* picture all sizes and sorts of men—immense variety of males from different ages, countries, doing different things like fighting, sporting, eating, dancing, smoking. Any illustrator needing a male symbol or just a man will find the motif here. 128pp. 8⅞ x 11¾.

HANDBOOK OF PICTORIAL SYMBOLS

Rudolph Modley and W. R. Myers

Convey ideas without words? 3250 visual symbols by government agencies, giant corporations, foremost designers for modern messages: trains, autos, road rules, women, factories, Olympic Games, tourism, do it, don't do it, etc. All copyright free except about 250, clearly indicated. Also decorative. 143pp. 8⅛ x 11.

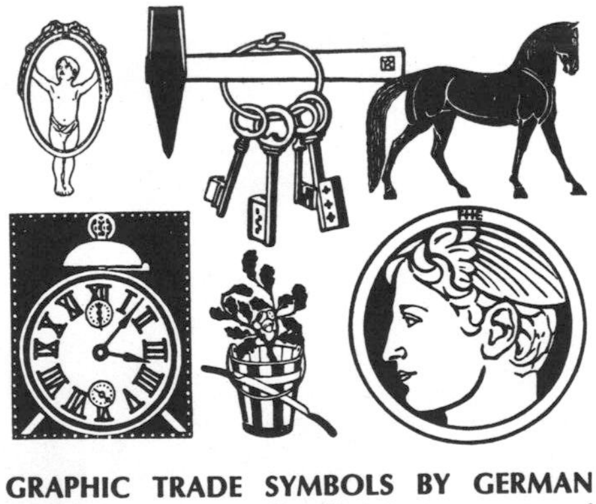

GRAPHIC TRADE SYMBOLS BY GERMAN DESIGNERS, F. H. Ehmcke. From renowned Klingspor Type Foundry of Offenbach am Main, 1907 catalogue. 293 beautifully designed pictorials for trades, activities, objects, institutions, plus 54 layouts, framed work, ads. Sports, agriculture, baking, beekeeping, ropemaking, bicycles, lithography, poison, bells, all in uniform style. Immediately readable for ads, logos, or decoration. 63pp. 8⅜ x 11¼.

HANDBOOK OF DESIGNS AND DEVICES, Clarence P. Hornung. 1836 sophisticated unit designs based on circles and circle segments, lines and bands, triangles, squares, rhomboids, pentagons, hexagons, scrolls, frets, loops, other geometric elements. Incredible variety, resources. Based on Japanese, Egyptian, Classical, Islamic originals. Eye-catching, suggesting both variety and geometric perfection. 260pp. 5⅜ x 8.

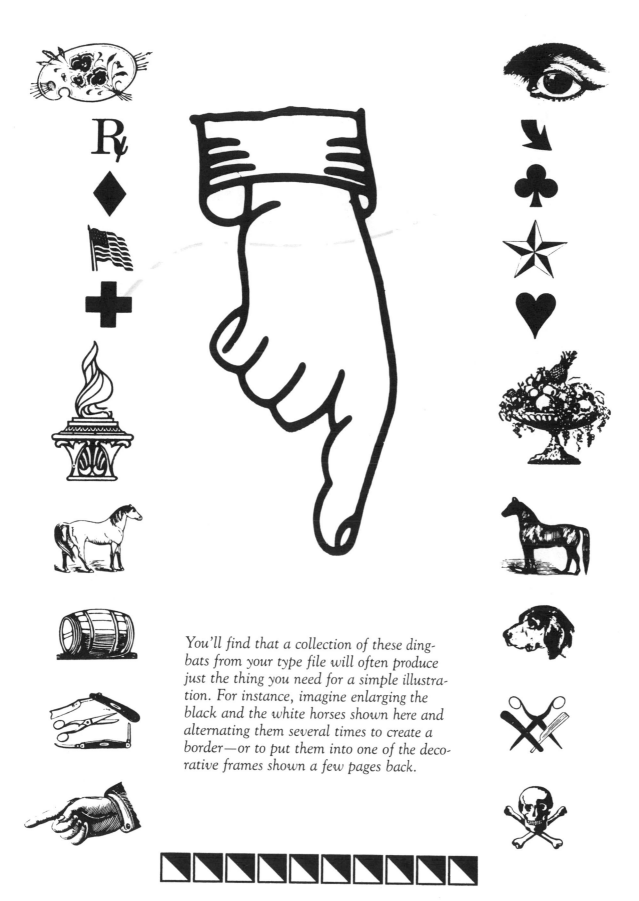

You'll find that a collection of these dingbats from your type file will often produce just the thing you need for a simple illustration. For instance, imagine enlarging the black and the white horses shown here and alternating them several times to create a border—or to put them into one of the decorative frames shown a few pages back.

H. MUNCE DESIGN

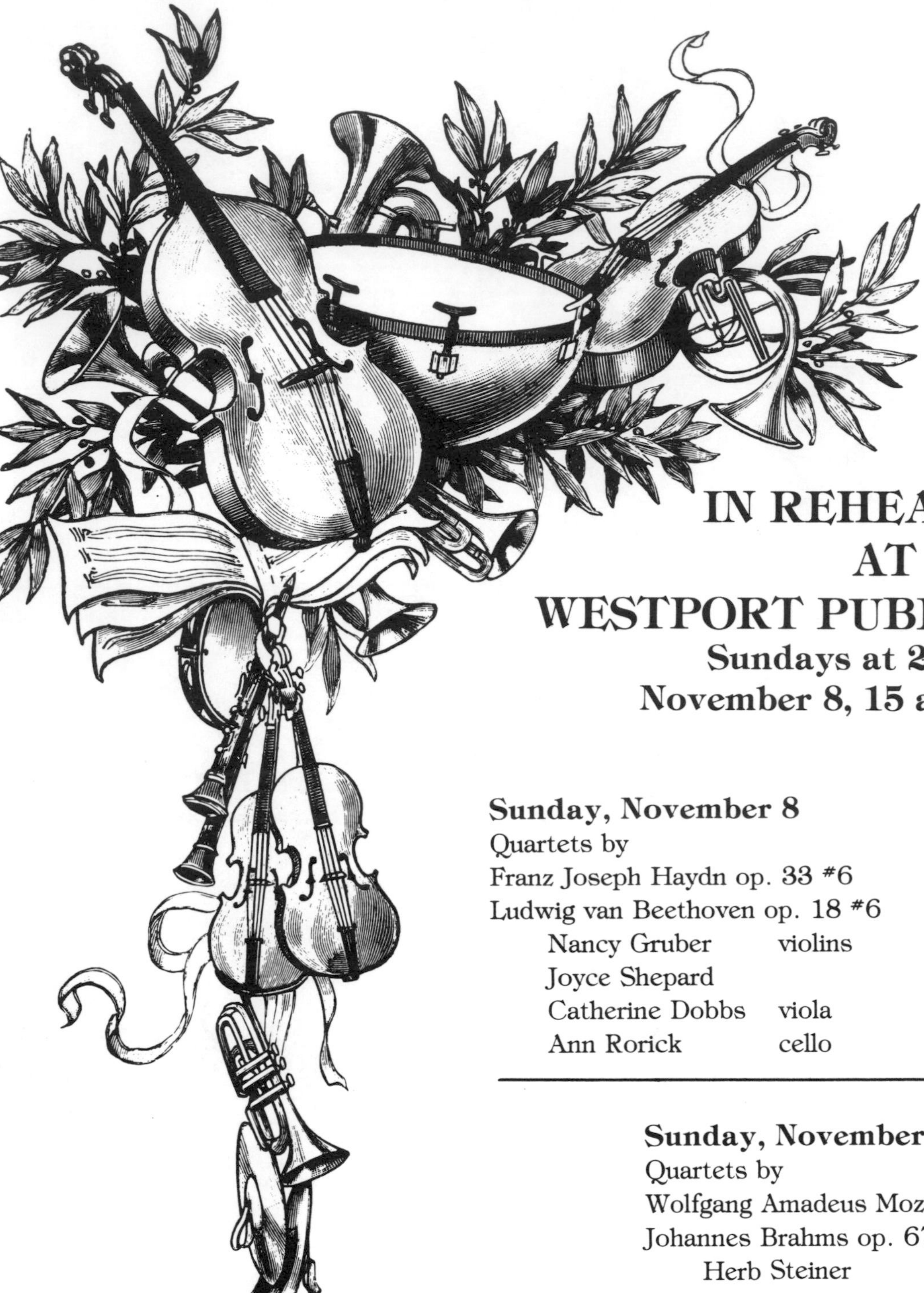

IN REHEARSAL AT WESTPORT PUBLIC LIBRARY

Sundays at 2:30 p.m.
November 8, 15 and 29, 1981

Sunday, November 8
Quartets by
Franz Joseph Haydn op. 33 #6
Ludwig van Beethoven op. 18 #6

Nancy Gruber	violins
Joyce Shepard	
Catherine Dobbs	viola
Ann Rorick	cello

Sunday, November 15
Quartets by
Wolfgang Amadeus Mozart K575
Johannes Brahms op. 67

Herb Steiner	1st violin
Joan Peach	2nd violin
Catherine Dobbs	viola
Paul Glenn	cello

Sunday, November 29
Quartets by
Wolfgang Amadeus Mozart K387
Johannes Brahms op. 51 #2

Robert Levine	1st violin
Nancy Gruber	2nd violin
Sam Gruber	viola
Jean Swain	cello

Drop in anytime.

Help yourself to refreshments.

This is an instance where I made use of an old steel engraving in the public domain. It's from an excellent book entitled "A Source Book of French Advertising Art," compiled by Irving Zucker and published by George Braziller, N.Y.

In addition to its hundreds of superb and exclusive drawings, it also has many pages of borders and shapes from which you can delete the French copy and insert your own.

From the
Picture Collection
of the
Westport
Public Library
Design: Munce
The enclosed pictures are for your use for one month. Please return them to the library in person. To preserve the pictures for other users and to avoid fines for damaged pictures, do not use staples, tacks or tapes on them. Thank you.
Westport Public Library, 19 Post Road East, Westport, CT 06880 • 203-227-8411

This is a large label that gets pasted on a sturdy library envelope.

The old engraving from one of the reprint books originally stressed only photography. But for my purpose here I needed a combination of art and photography symbols. So I added the "art" by pasting in a line drawing of the paint box from an old supply catalog and made the lines heavier to conform with the original.

Here's an example of using a baroque decorative framed border and tailor-making it to serve one's purpose. This is how it looked when I revamped it for an article which appeared in Olin Magazine.

A Brief Guide to the Wines of the Piedmont Region

The following wines are all imported into the United States, and should sell for no more than $3 or $4 per bottle.

ASTI SPUMANTE—The hills around Asti are honeycombed with caves dug out for bottles of wines double-fermenting into sparkling wine. Asti Spumante is a delicate, sweetish, aromatic sparkling wine derived from the Moscato grape. Generally considered too sweet for mealtime consumption, it is delicious with fruit or sweets after a meal, or at parties.

BARBARESCO—Nebbiolo grapes produce this wine slightly less powerful and developed than Barolo, but dry enough to pucker your lips. Differences in elevation and soil composition around the town of Barbaresco account for the lighter quality, although this is still a robust wine. It should age for about three or four years. Serve with red meat, pasta, strong cheeses and game.

BARBERA d'ASTI—Barbera is the most common grape in Piedmont and produces red wines that vary widely in quality. Harsh when young, Barbera mellows with age, some very good years being 1947, 1950 and 1955. It is a rugged, full-bodied wine to be served with spaghetti and red meats.

BAROLO—The king of Italian red wines, Barolo is made from the Nebbiolo grape. It is deep red in color, full and fragrant. A strong, dry wine, but with a velvety texture, Barolo should age for six or seven years, when it takes on the unmistakable scent of violets. Serve with red meat, pasta and game.

FREISA—Freisa is a fruity, light red wine and one of the most charming of the Piedmont's minor wines. It sometimes carries a hint of sweetness; enough to characterize it as semi-dry, but not enough to merit the description "amabile," or "slightly sweet" on the label. Some Freisas also have a bit of effervescence, with "frizzante" on the label. Serve with veal, pasta and meats with sauces.

GATTINARA—The village of Gattinara produces less than neighboring villages, so this wine may be harder to find. It is a dry, velvety red wine with a balanced, medium body, lighter than Barbaresco. Gattinara should be at least four or five years old before being drunk, but it will still show a remarkable delicacy.

GRIGNOLINO—This is the lightest of the major Piedmontese red wines, produced from the grape of the same name. Despite its delicacy, it is quite dry, with a faint bitterness that lingers in the aftertaste. Its pale red color gives away this wine's lighter density and body compared to the generally deep red Piedmontese wines. Grignolino peaks young, and declines after four years.

NEBBIOLO—Made from the excellent Nebbiolo grape outside the delimited regions of Barolo, Barbaresco and Gattinara, this is a little lighter than its three illustrious cousins, but still a hardy red wine with a sturdy aroma. Wine labeled Nebbiolo may be either dry (secco) and still, or slightly sweet (amabile) and sparkling (frizzante or spumante), so check the label carefully for the one you want.

This is a mailing piece for an exhibition of two wildlife artists.

These drawings come from a public domain publication that had collected line cuts from old dictionaries. Each had the animal's name set in type, so I left it. I also left the crooked arrangement untouched.

Below are the same creatures again—used this time to illustrate a newspaper review for a book about Noah's ark for the Greenwich Time. As you can see, I needed two of each this time, so I xeroxed another set and pasted them in pairs. I drew the cartoon of Noah and the Ark and added the skunks on the gangplank and the sloth hanging from the tiller. The total effect was an attempt to decorate a newspaper page that is usually either a sea of gray type or is illustrated only by regular cartoonists's panels.

Joe Alex Morris **Book Review**

Noah's ark: Getting to the very bottom

SURVIVING THE FLOOD. By Stephen Minot. Atheneum, New York. $14.95.

One of the most clamorous and controversial developments in recent American society has been the interest of millions of persons in a revival of religious fundamentalism. Television and radio evangelists have thundered the old time gospel and collected vast sums from enthusiastic converts. The Moral Majority has set out to eliminate unbelievers from political office. And in Arkansas the courts have been called on to rule that the creation theory — that is, the Biblical account of the Garden of Eden — is just as scientific, if not more so, than the Darwinian theory of evolution.

There have always been certain unanswerable questions raised by the evolution theory (the missing link, for instance) but the so-called "creation science" has been faced with the problem of filling in numerous gaps in the Biblical history of mankind from the time of Adam's bad luck with the apple on down through the ages. Some of the gaps concern Noah and his ark, for which fundamentalists have searched Mount Arat in recent years with questionable results, or no results at all.

This has not intimidated Stephen Minot, a Connecti-

For its first exhibition of the 1981 season the GE Gallery presents the work of two of the country's leading wildlife artists: John Hamberger and George Luther Schelling.

John Hamberger is the author illustrator of several children's books. He has also illustrated over forty-five books on nature and wildlife. Mr. Hamberger's illustrations have appeared in magazines and periodicals and his paintings have been shown in galleries across the United States.

The Museum of Natural History in New York City commissioned him to paint the "Life Cycle of the Coho Salmon" for the Hall of Ocean Life. He has also been commissioned by the National Wildlife Federation to paint birds and animals for National Wildlife stamps, Christmas cards and letterheads.

Among Mr. Hamberger's awards are: the Philadelphia Book Show Award, the Christopher Award, the Outstanding Science Book Award, and the Design Publication Award. As a recipient of the Christopher Award, he was selected to present this year's award to five children's book illustrators.

George Luther Schelling of Laceyville, Pennsylvania has exhibited his paintings at the following galleries: Grand Central Art Galleries, The Kennedy Galleries, The Crossroads of Sport, Abercrombie and Fitch, the Smithsonian Institute, the Brandywine Museum, the Royal Ontario Museum, Madison Square Garden, The Allied Artist of America, The National Arts Club, the National Art Museum of Sport, The Art Emporium, Husberg Fine Arts Gallery and the Jackson Hole Art Gallery.

His illustrations have appeared in the following publications: Outdoor Life, Sports Afield, Field and Stream, Audubon, The Garcia Corporation, Boys Life, Animal Kingdom, The Pennsylvania Angler, Ranger Rick, Daiwa Fishing Annual, South Carolina Wildlife, Georgia Sportsman, Seagram Sportsman Calendar, and the Hartford Life Calendar.

Mr. Schelling is a member of the Society of Animal Artists, and the Society of Illustrators.

aboard the ark, adding that anybody who studies the Biblical begats and ages will realize that the old man was still alive when the flood came and that he dies in the ark and was buried at sea. He also explains how the prehistoric great sloth disappeared: there was only one in the ark and it was accidentally killed and eaten for supper. Goodbye great sloth!

But those are mere tidbits in Ham's account of how difficult it was to collect all the animals (elephants were to big to get aboard and were abandoned) and what an unbelieveable mess it was inside the ark after all the hatches and portholes were closed and shuttered and after hundreds of men, women and children were beaten off when they tried to cling to the deck railing as the waters rose.

This may make Ham's memoir sound grim but on the contrary it is lively and interesting, often very witty and laced with a love story that is both hilarious and tender. Noah is a huge, hard-drinking, roaring kind of man and his family is rowdy, quarrelsome and so accustomed to running things that they take all their servants on the ark. Ham's brothers, Shem and Japeth, have brought a huge supply of wine aboard. Among their wives' servants is a beautiful foreign slave girl, Sapphira, who is equally skillful at taking care of a sow birthing piglets on the lowest deck and of Ham's amorous propensity on any deck.

The adventures of Ham and Sapphira and their shipmates make a rollicking, lusty tale enhanced by a leak that threatens to sink the ark, by a rebellion of the servants and by the fact that Noah angrily refuses to open the portholes and let anybody see what is out there — even after the rain stops. This mystery haunts Ham until, unknown to his father, he gets a glimpse through a porthole -- and is shocked by the realization that Noah is conducting a fantastic stonewalling operation to make certain that history is written the way he wants it written.

Tensions within Noah's troublesome family burst into violence when the ark finally settles on land and Ham, more sensitive than the others, cannot reconcile himself to acceptance of the official story of their voyage. He has matured and has begun to seek answers to puzzling questions, particularly why was Noah's family saved when the rest of the world was engulfed in the flood? Or why was Sapphira born to be a slave?

Ham's story of how his father lay drunk and naked in his tent and how in a terrible rage he cursed Ham and all Ham's descendants differs considerably from the official fundamentalist version but the net result is much the same. Soon Ham and Sapphira took off for exile to found a family and raise four sons who built "some dazzling cities."

"Oh," the memoir concludes some nine centuries after the flood, "I know what they say about Sodom, and there are some who have harsh words for Gomorrah and my adopted home, Babylon, but these cities are impressive. They have energy. They will be remembered."

———

This book is available at Greenwich Library and the Perrot Library, Old Greenwich.

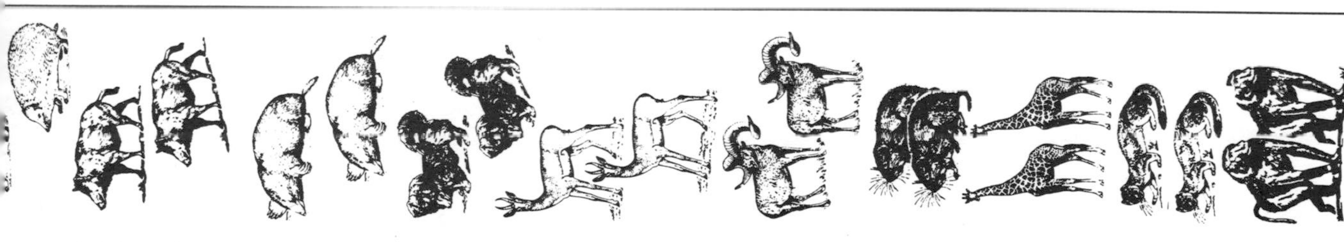

For a children's announcement, I used this charming drawing that my daughter did when she was five years old.

These drawings are from the sketchbook of Eve Donovan, a college art major.

The one on the left has been published in the Journal of Traditional Acupuncture.

If you can acquire original art work from superior students, you'll discover that it can have a fresh feeling that more practiced hands often lose.

This is a page I've gotten a great deal of mileage out of. The use of time as a symbol comes up often.

Banish the time eaters... and get two hours more work done every day!

Studies show that almost everyone wastes two hours every day...that's 30 days a year! Not because they want to...or because they're lazy. But because they're plagued by time eaters!

This meeting shows you how to cut down on wasted time ...kickout the time eaters. You find out how to *control* your time to fit your own plans and job needs.

Who says you can't break a habit?
Time use is a habit that can be altered. You'll find out how to keep track of your time at this course. And you'll see what percentage of your time is spent on things you must do ...and where your time-use priorities lie. It's a big step forward in putting time to work for you.

Help others use their time more productively
As a manager, you're using other people's time. Your secretary's. Your staff's. What can you do to schedule their work time better? Find out at this meeting. And take a look at how you spend time as a member of the group. For instance—see what you can do to create successful conclusions to meetings.

Don't waste another minute!
Get your registration in right now! Use the Registration Card on this brochure. Or, for faster service, phone the Registrar in New York City at (212) 246-0800.
Ask about in-house training! For details about related in-house versions of this program, contact AMA's In-House Development and Training Division in New York. (212) 586-8100 Ext. 162.

▲

For a mailing piece for a lecture series about some far out literature, I chose the bizarre-looking fellow above You can see how I treated him on the opposite page. I was printed in purple to heighten the dramatic effect.

The art on the left was made by joining two engravings. You can play endless graphic tricks by combinin elements.

H. Munce, Design

MAD, MAD, MAD, MAD WORLD!

WESTPORT PUBLIC LIBRARY

TUESDAY, October 6, 1981, 8:00 p.m.
THE MOVIE GOER by Walker Percy, 1961
Recipient of the National Book Award, Percy's first novel depicts the adventures of a man whose real life exists for him on the screen. It will lead off our discussion of illusion versus reality.

TUESDAY, October 27, 1981, 8:00 p.m.
THE WORLD ACCORDING TO GARP by John Irving, 1976
An outrageous, passionate novel about a writer (and, therefore, about fantasy versus reality,) this book raises all the important questions about absurdity and realism, humor and seriousness.

TUESDAY, November 17, 1981, 8:00 p.m.
BIRDY by William Wharton, 1978
Birdy is a fantastic, heroic character whose dreams of flying both imprison him and set him free.

TUESDAY, December 8, 1981, 8:00 p.m.
A CONFEDERACY OF DUNCES by John Kennedy Toole, 1980
Published posthumously (and with the efforts of Walker Percy, the first novelist in this series,) this hilarious, outlandish book won this year's Pulitzer Prize. Like all the books to be discussed, its hero is a visionary of sorts and its humor is meant to be taken seriously.

Paperback copies of the above novels are available at local bookstores.

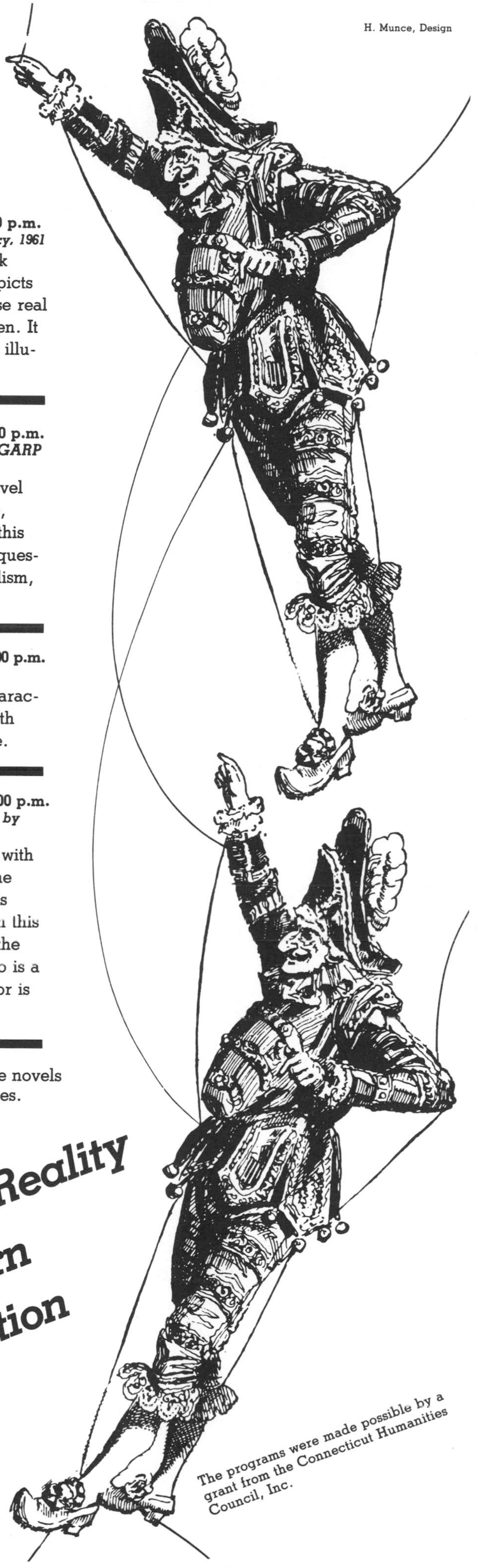

Fantasy and Reality in Modern American Fiction

The programs were made possible by a grant from the Connecticut Humanities Council, Inc.

There are clip art services you can subscribe to. Their catalogs offer you dozens of ready-made drawings to choose from. There are situations for all holidays and varied events—sports, business, etc. The styles vary from realistic to decorative to cartoons.

These examples are from the Volk Company in Pleasantville, New Jersey.

33

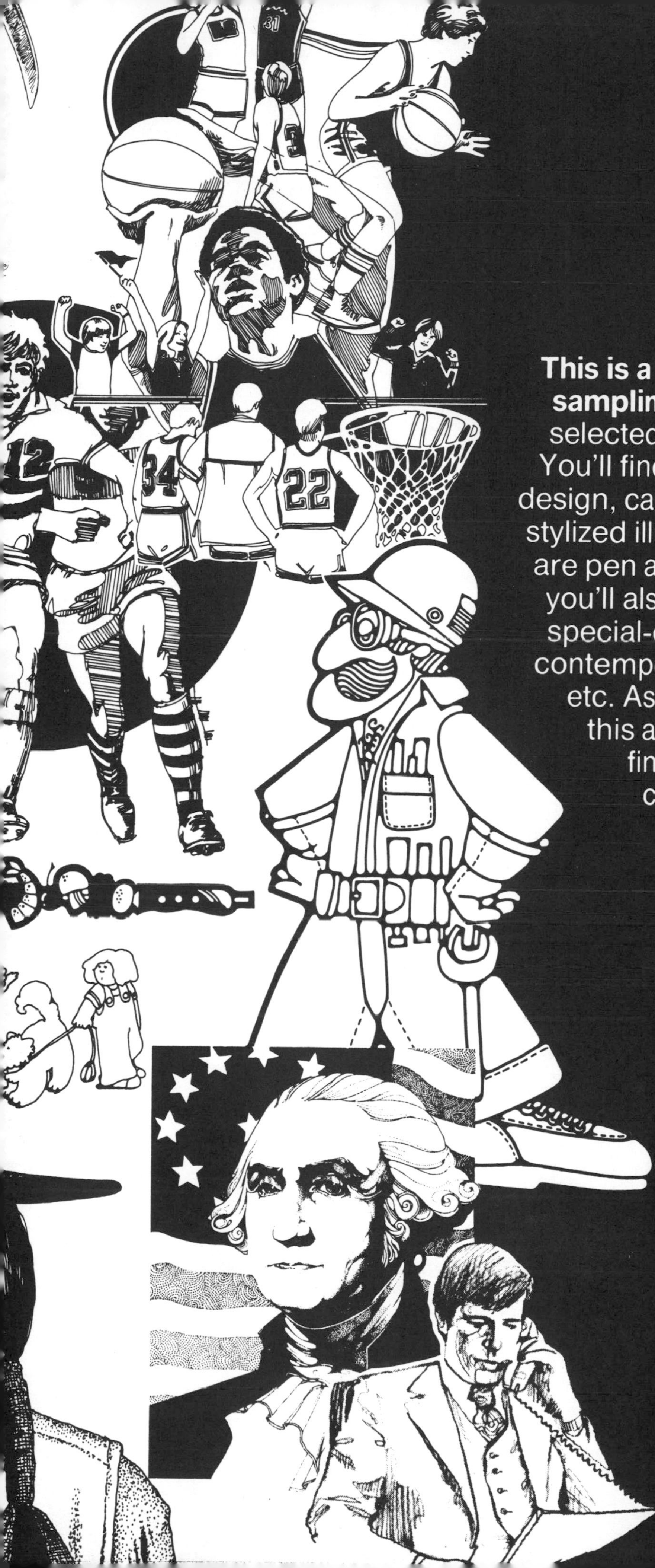

This is a representative sampling of Clipper® art selected from recent issues. You'll find representational, design, cartoon, realistic and stylized illustrations. While many are pen and ink renderings, you'll also notice scratchboard, special-effect photographs, contemporary hand-lettering, etc. As in every CLIPPER issue, this art was created by the finest professionals in commercial illustration.

Dynamic Graphics Inc.
Clipper Creative Art Service,
Peoria, Illinois 61614.

PHOTOGRAPHY:

A myriad of hints, suggestions & instructions by a snapshot shooter on how to produce non-pedestrian & undull pictures for whatever purpose.

Odds are that you will more often deal with photography than with art work for your assignments.

This happens because your client is likely to supply you with photos that have been gathered for the purpose. Most often they're portraits of people; officers, committee members, etc. And too often they've been taken by different photographers, so you'll find no consistency in the lighting. But, most distressing, such business photos have the banal look of sterile yearbook pictures—stiff, posed and stuffed.

When you're stuck with such photos, you have two ways of ridding them of some of their ordinariness; by bold cropping, and by the use of patterned printing techniques, both of which will be dealt with later in this section.

The ideal, of course, is to be in charge of *having* the pictures taken, or in taking them yourself if you're proficient in photography. In either case, you can be sure the pictures will have a unity, a more candid quality and a bold simplicity that will insure good reproduction.

Often when you inherit photos that were taken at affairs or meetings you'll find that they are stark prints, brilliantly lit in front, but with harsh black shadows elsewhere. Also, this kind of "News" picture is often shot hastily to capture the people and the event, but with little thought of the jumbled and disorganized background.

When you're forced to deal with this kind of photo, retouching is often called for. That subject will be discussed later, too.

If you're put in charge of producing photographs, you are now an Art Director, and your duties will be the same as any highly-paid A.D. from an advertising agency or a big time magazine. It becomes your responsibility to select a photographer—one you can afford and one whose work has the flavor you favor for your assignment.

If you work in a fairly large town or city, there will be several photographers listed in the Yellow Pages. Call them and arrange to see the kinds of work they do. After you've agreed on one, it's up to you to express your feelings about the look you're after. Then you should be explicit about every aspect of the shooting, except how to work the camera.

If you have lighting preferences, express them. Have an opinion, too, about clothes, props and location. If models are needed, you'll be shown 8 x 10 inch composites to choose from.

In the case of childrens' model pictures, be sure to ask when they were taken. Kids change radically in months.

It's important to know in your mind's eye how you *think* you'd like the pictures to look. But know that when you get to your actual location all sorts of ideas will begin to materialize based on the place—things you couldn't have thought of: interesting backgrounds, places to sit or stand on, etc. This is the point where you and the photographer can get your kicks. Swap ideas and enthusiasms with him. Say what you think. But don't inhibit him from taking the picture as he sees it—so long as it conforms to your needs.

I've often been guilty of stepping on photographers' toes with my wants. But, I learned to control my eagerness by first having them shoot what I wanted, then letting them cover it as they saw fit. I have no hangups about later using the best shot, no matter whose idea it was.

Group shots present a special problem; perhaps because they're so common in local newspapers, fraternal publications and trade journals. Do your best to avoid the usual even rows of people. There is often something unique in the place you're shooting to allow you to compose groups in more interesting ways.

Now you must be part art director and part movie director to get what you want. Models need to be told what to do. Good ones with intelligence and ability will lend more than their presence to the situation. Avoid dual directions. Either you talk to the models or let the photographer do it after you've made your wishes clear. The latter method is usually the best for all concerned.

In addition to portraits and group shots, circumstances will sometimes call for you to devise and supervise some other categories of photography—the staged shot and still lifes.

Staged pictures can be done in the studio or out of doors. Subject matter and weather will usually determine which is preferable.

The still life is just what its name implies. The object can be anything, literally from soup to nuts. . . and bolts. Your job, along with the photographer's, will be to arrange and compose the props to make a visual statement.

Lighting is very important in still life work. Here your photographer will supply the professional savvy.

One of your greatest responsibilities in dealing with photographs is to first see that your picture fits the space and dimensions of your layout. (Don't shoot a vertical picture for a horizontal area.)

Many photographers like to take preliminary Polaroid shots so you both can check out trouble spots before the actual shooting. Do this whenever you can. It's a very reassuring aid for everybody concerned.

This still life had to say three things: filed pictorial material, research books and a portrait of the woman who instituted the exceptional picture file in the Westport, Connecticut public library.

Bill Noyes, the photographer, first enlarged a small photo of Mrs. Sherwood to fit the antique frame which was chosen for its period look and for its appropriate scale to the file drawer. I used the frame by itself for the back of the folder.

film is your SUNDAY PASSPORT to 8 great places...

MEXICO	ITALY	GREAT BRITAIN	GERMANY & AUSTRIA
Mexico City, Mayan ruins, villages, traditional arts and crafts. OCT. 14, ■ 4-6 p.m.	Naples, Pompeii, the Alps, and Italian art, music and history. OCT. 21, ■ 4-6 p.m.	London and Scotland, the literature and history, and the enduring beauty of the countryside. OCT. 28, ■ 4-6 p.m.	The music, the Danube, the Rhine, the mountains, and old cities and new. NOV. 4, ■ 4-6 p.m.
SPAIN Goya, beautiful villages, the island of Majorca, the food, history and culture. NOV. 11, ■ 4-6 p.m.	**GREECE & TURKEY** Greek festivals, Istanbul, Byzantine art and ancient history. NOV. 18, ■ 4-6 p.m.	All at Coleytown Junior High School Westport. Sundays, October 14 through November 18, 1973. Series fee for Westport residents: $2.00 Co-sponsored by Westport Public Library Westport Division of Continuing Education	

The props in this simple still life all say "travel" and "movie." They were chosen for their quick identity and their various shapes — circular, rectangular, vertical and horizontal.

Note Bill Noyes' nice touch in letting a bit of film curl out from the reel, and allowing light to come through the frames and the sprocket holes.

Bill Joli

HAROLD VON SCHMIDT

To be sure, it's not every day that you get the opportunity to shoot such a wonderful character as the legendary and picturesque Harold Von Schmidt. But everyone from an office manager to the chairperson of the Woman's Club has their own individual look, and possible interesting setting. It's up to you and the photographer not to settle for an obvious trite solution. Dig for an original way to portray your subject.

CAL SACKS

Here is one made right in the artist's back yard at the river's edge. It's there that he fishes out some of the objects he uses in his constructions and as material for his antiqued signs. Note the rightness of everything—the background, the water, the subjects, work clothes; all appropriate for a man who works with junk and tools.

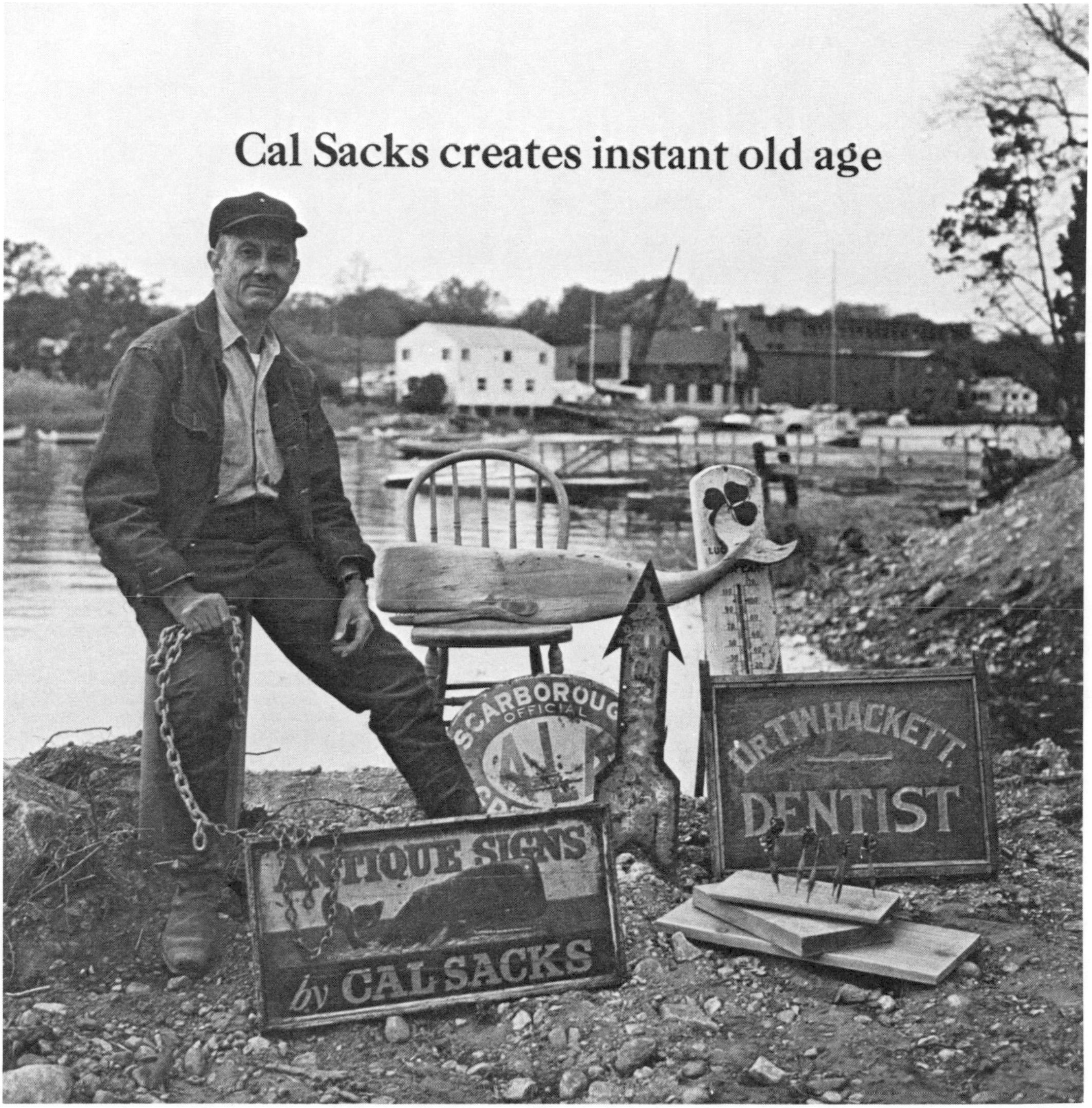

Bill Joli

STEVAN DOHANOS

The title of an article in North Light Magazine brought this setting about.

Wearing the house painter's cap was Dohanos' idea—a lovely touch. Unhappily, this grand old house was later bulldozed to the ground. It's good to have this record of it.

Don't paint your house–
I'll paint it

Bill Joli

DEAN ELLIS

This artist, a man who does extensive research for his assignments, was appropriately photographed outside the New York Public Library against this marvelously patined base of one of the flagpoles.

Bill Joli

BETTY FRAZER

This artist specializes in subjects in nature—mostly animals. Since she's a New Yorker, I wanted to keep her there—so we went to the children's zoo in Central Park and photographed her with many creatures. But this one of the show-boating rooster won hands down.

Bill Joli

HARRY CARTER

For a portrait of watercolorist Harry Carter, this location was chosen—and with good reason: Carter had long guided an exhibition program in this picturesque building, the Rye, New York Public Library. Observe what a role the great rococo iron bench plays in this picture. A good eye and some good luck will turn up such finds. Often you can base an entire composition on something as interesting as this.

Bill Joli

WALTER BROOKS

Here is an example of photographing a person in a location germane to his profession or interest. Walter Brooks is a versatile artist who specializes in woodcuts.

It was a simple matter to get permission to do the shot at a lumber yard. But not just any lumber yard. In an area where there were several choices, I think you'll agree that the inclusion of the old ivy-covered factory in the background made this an ideal one for a New England artist.

Bill Joli

FAIRFIELD WATERCOLOR GROUP

Group shots can be difficult because of the need to catch a good expression on everybody. Always be sure to shoot exhaustively. The cost of film is nothing compared to reshooting. Beyond that, you must look for a location that is both appropriate and photogenic. I think this shot has everything: ten good portraits—and a fascinating studio that you can study and keep finding new fascinating details.

Bill Joli

BOB GEISSMANN

There were two reasons to photograph this designer in front of the U.N. Building. One, he was a confirmed New Yorker and his studio was in the neighborhood, but more importantly, his article dwelt on Piero della Francesca's painting, "The Battle of Constantine," with its raised lances and banners.

I saw a parallel between them and the poles with the flags of the nations at the U.N. So we just walked down to First Avenue, set up an easel and examples of his work and shot. Such goings-on don't faze blasé New Yorkers.

Bill Joli

WALTER and NAIAD EINSEL

A handsome double portrait of a husband and wife team surrounded by his work, their interests and their beautiful home.

Bill Joli

Bill Joli

MARILYN HAFNER

Marilyn Hafner is an artist who makes wonderfully whimsical drawings—this quality carries over into everything else she does. People love to receive mail from her because she decorates her letters—and the envelopes.

So, for this portrait we had her stride past this photogenic mailbox with her arms loaded with some of her graphic goodies. What a lucky find to have the mailbox next to a street planter, and to have those unusual lamps in the background.

Bill Noyes

JOHN McDERMOTT

The illustrator-writer John McDermott, in addition to his other talents, was a moviemaker and Civil War buff, so this quaint monument was a perfect background for him—and the church in the background was another great break; it was appropriate, excellent in shape and perfect in color. It's brown, the monument is green and McDermott is wearing a red shirt—all this against a cloudless blue sky. Perfecto. Note how the composition builds—from Mac up to the soldier, then up to the steeple.

SOME SIMPLE TECHNIQUES:

Here are some graphic aids that will allow you to decorate and enhance your work.

These are a few inexpensive and effective methods of giving the illusion of shadow, depth and tone to drawings and stark areas in need of softening, variation or separation.

They are easily applied. Lay a piece over the selected area, press down lightly, and carefully cut exactly on the show-through line. (An X-acto blade lightly wielded is best for this purpose.) When you're done, lift away the excess and press the remainder more firmly in place. The warmth of your thumb will make the waxed surface adhere.

As you can see in the accompanying examples, these tone and shading patterns come in endless varieties.

A note of caution concerning the dot screens: if your art work is to be reduced or enlarged, it will naturally affect the closeness of the dots—and you must allow for that. You can see the problem if you start with a screen in which the dots are already close and then compact them even further, they'll clot and produce ugly solids when printed.

Anything you see here in black and white can be reproduced in color. A background of these patterns printed in color with a darker solid object over it is very effective.

'his is what happens when the reduction of close ›atterns is too great. ►

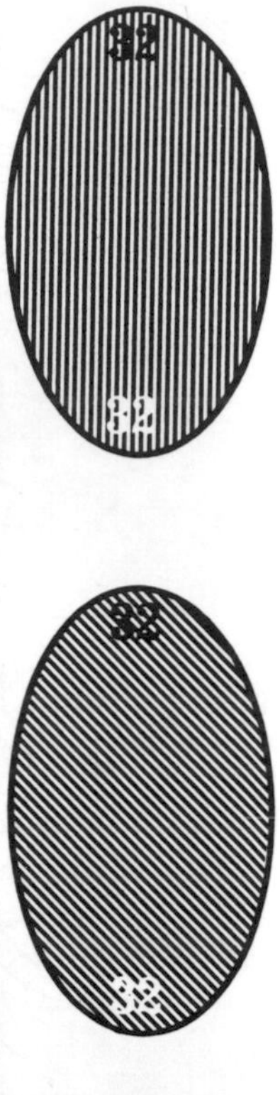

Hands Off!

Several different patterns are used in these three drawings. As you can see, there are a myriad of others to choose from; these pages give you just an inkling.

Formatt

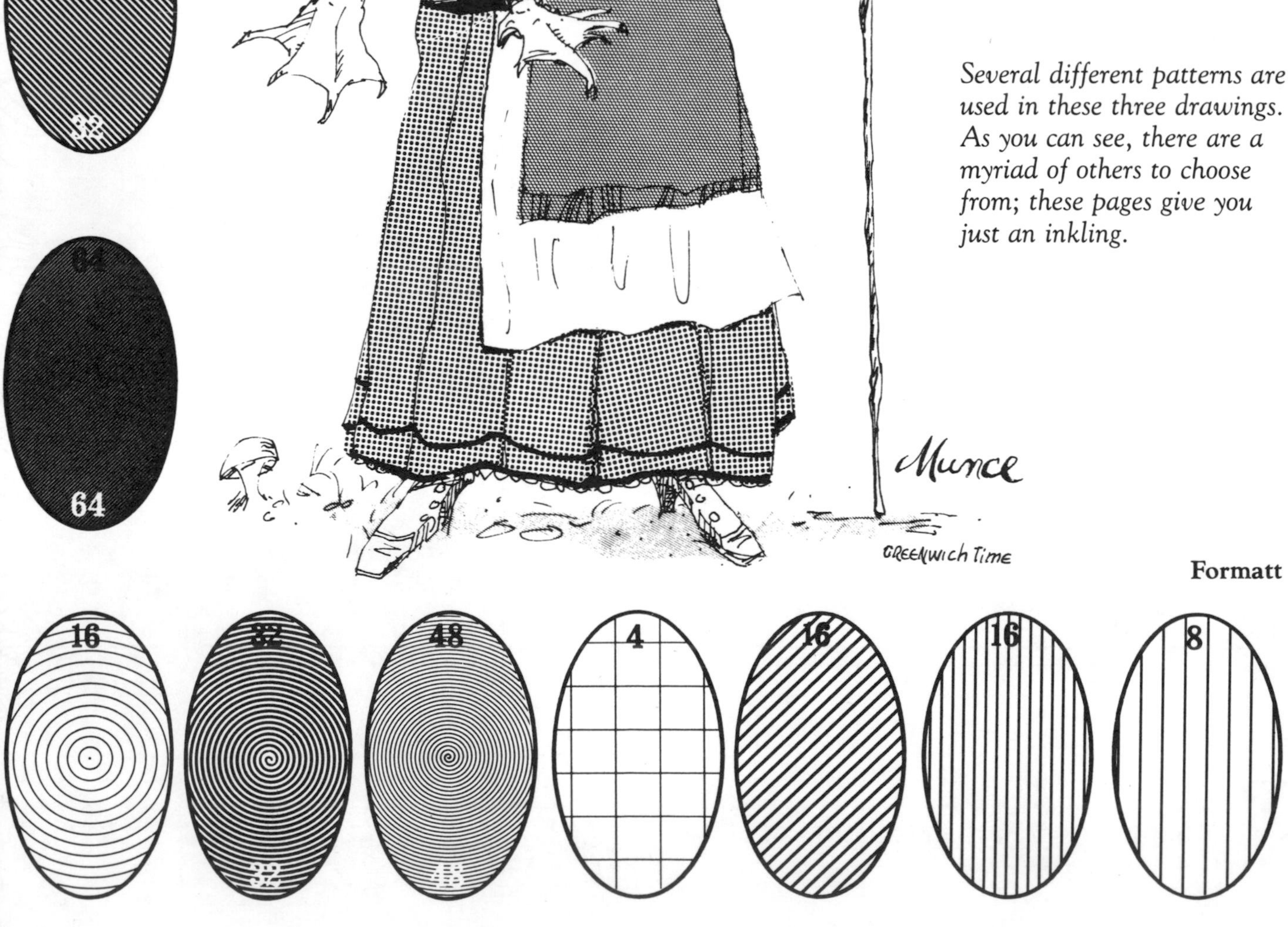

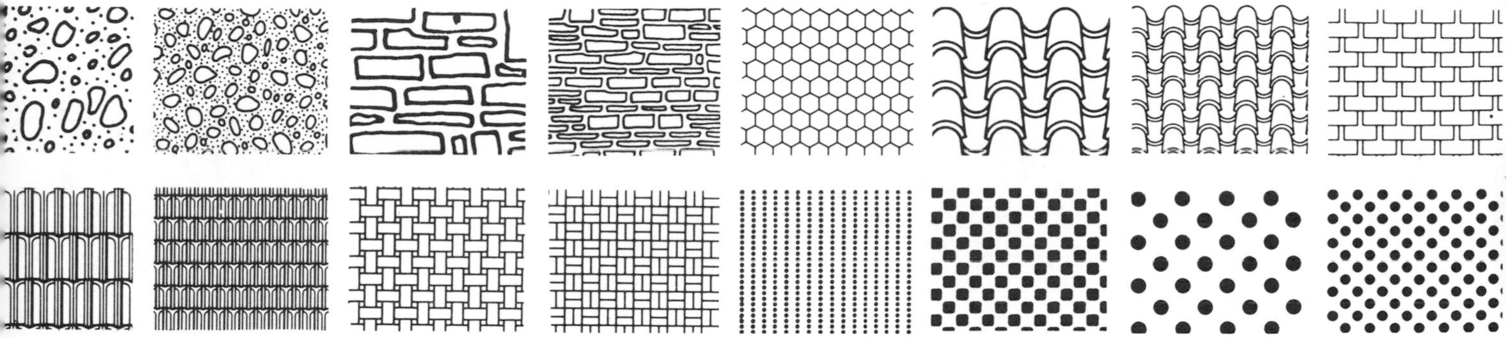

Letratone

MEZZOTINT

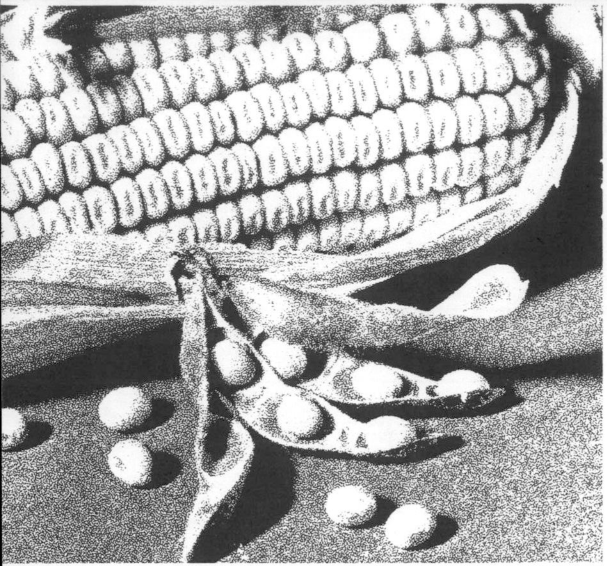

SPECIAL SCREENS

TONELINE MEZZOTINT /

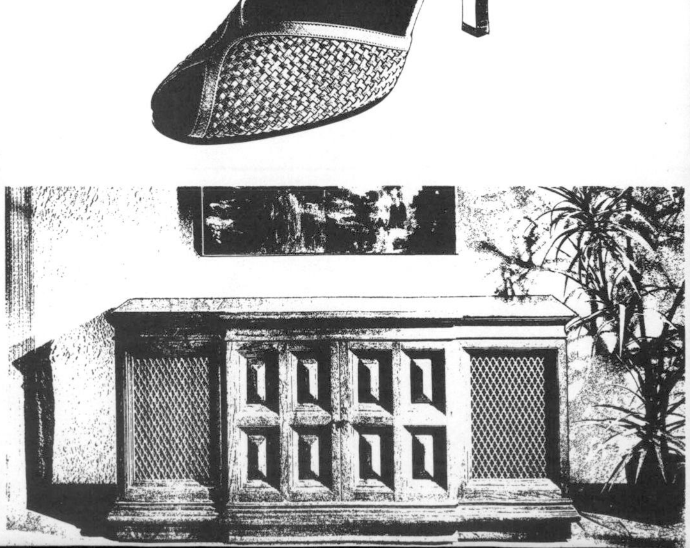

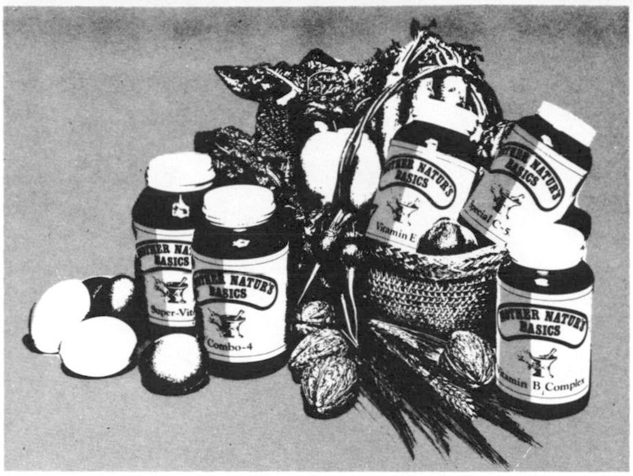

These are examples of mezzotint patterns — an "art" treatment you can apply to photographs. The effect is still photographic, but as you can see, the process lends a new look to the photo.

From the Mask-O-Neg catalog. The company is in New York City.

METRO ASSOCIATED SERVICES
33 WEST 34th STREET,
NEW YORK, NY 10001 (212) 947-5100

STRAIGHT LINE

RETOUCHING:

It is erroneously said that you can't make a silk purse out of a sow's ear. A good retoucher can.

Photo-retouchers are skilled artists who specialize in the repair and enhancement of photographs. By the use of paint brush, air brush, dyes and bleaches, they perform amazing feats. They can strengthen weak areas, take out unwanted objects, remove blemishes, double chins, dark circles and so on. When you need the services of a retoucher, seek out a good one, because overdone, obvious retouching can be worse than none at all. The good ones know just the degree needed to work well with the kind of reproduction it will be subjected to.

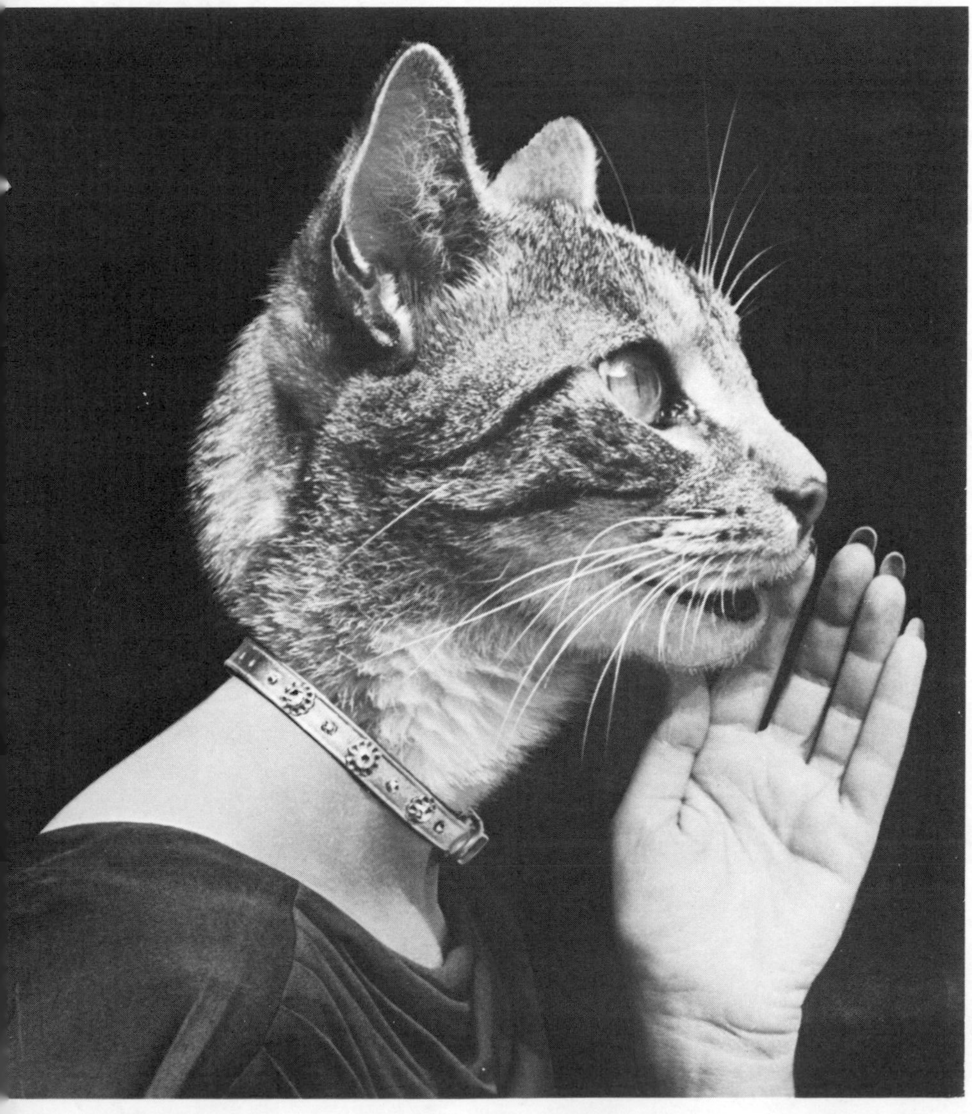

These examples of excellent professional retouching are the work of Bernie Burroughs of Westport, Connecticut.

Over the years he has encountered every possible kind of problem; in the human figure, in soft fabrics and in hard demanding mechanical objects. A cross section is shown here.

This is an example of the retoucher on busman's holiday—doing a job for himsel[f]. You figure out how he did it.

This is a fool-the-eye performance. Burroughs lettered the words on a color print to give the illusion that they were carved in the rock. Each letter had to have depth, and each had to conform to the rock's irregularity.

OF MAKING
MANY BOOKS
THERE IS NO END...
ECCLESIASTES XII·12

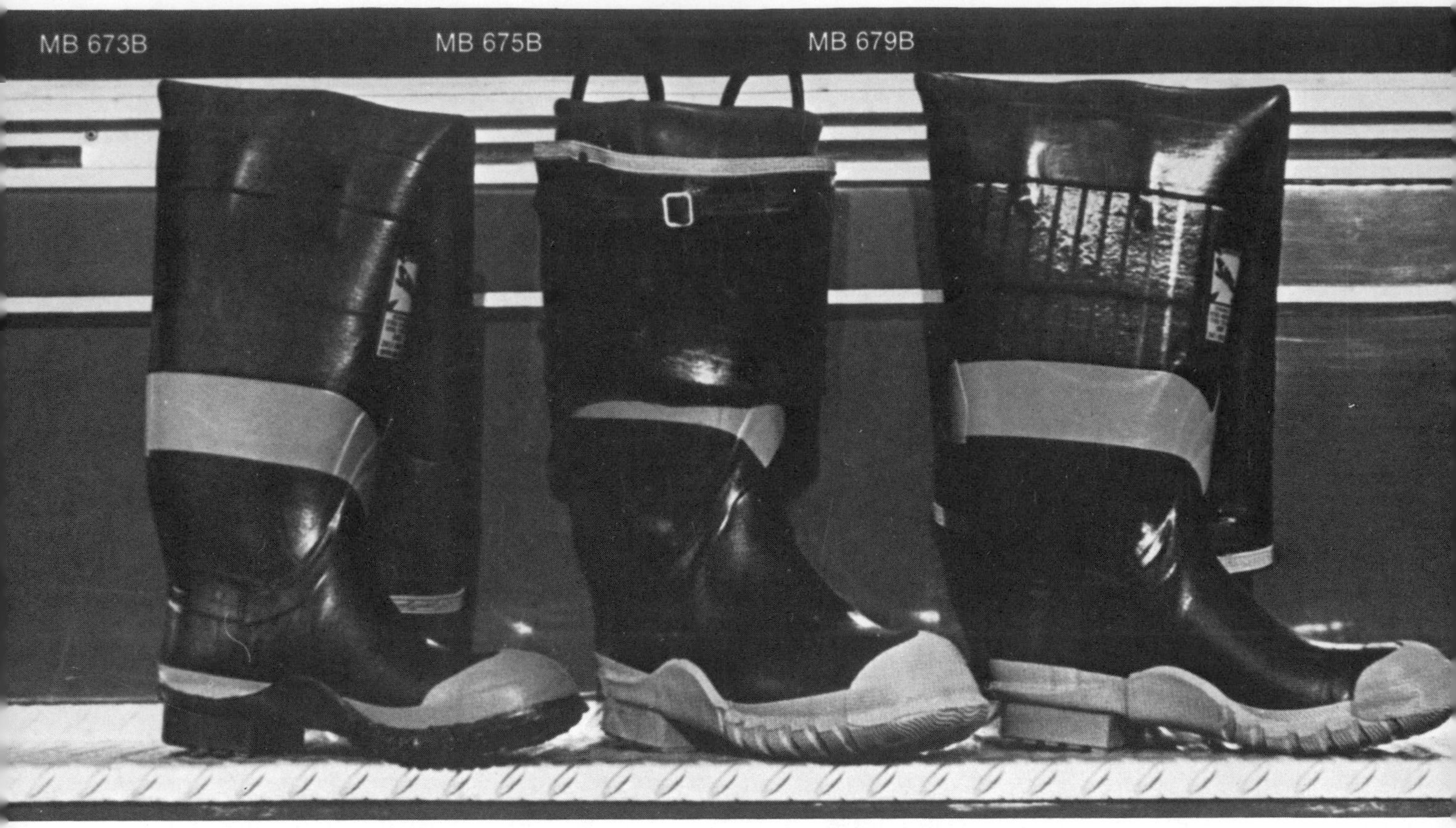

UNIROYAL

The center boot had to be removed from the top photo. Below, the same shot with the boot gone and the left and right ones moved next to each other. In the meantime, he brightened up the fire engine. This was originally done on color prints. It entailed the use of blade, paint brush and air brush.

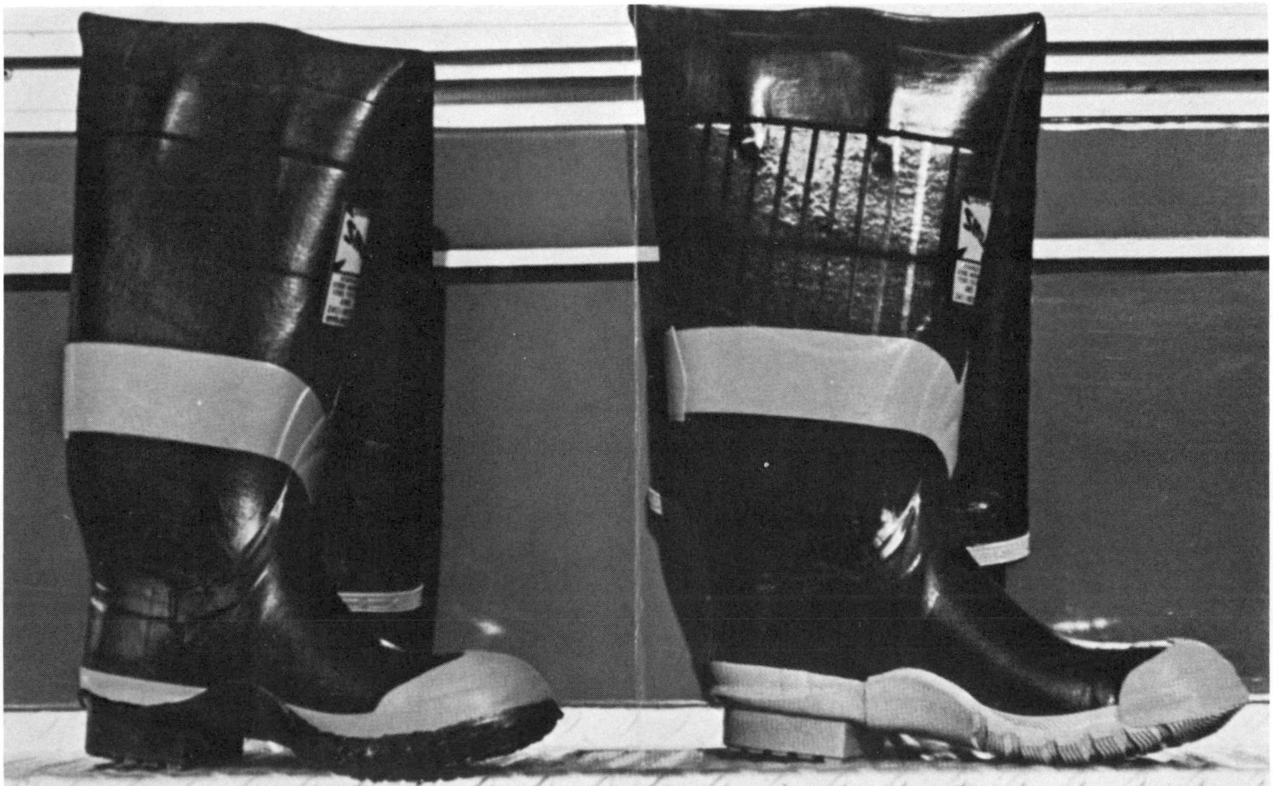

Photographing shiny metal objects gives photographers fits, because they act as mirrors and pick up unwanted reflections, especially when they are as smooth and polished as this one.

The first concern is to get rid of the unsightly ones, and to do it without destroying the form of the original. Changes were also made in the background and the base.

Here's the sparkling result!

This is a complicated photo—full of faults, and greatly in need of strengthening in some places—elimination of unwanted objects in some places and redrawing in others.

Place a sheet of acetate over it, and with a grease pencil, practice marking your corrections for the retoucher in the simplest possible way.

Follow the edge all around to separate the figure from the back-ground

This is a simple retouching job you can learn to do yourself with only a small brush and white paint.

First mount the print on cardboard and then paint up to your subject. Be very careful not to nick into it.

When you are ready to step up to more advanced studio and design work, a number of fine books await your study. Here are a few.

ARAPAHOE COMMUNITY COLLEGE
3 1717 00040 4709